CAPILLARY LIQUID CHROMATOGRAPHY

MACROMOLECULAR COMPOUNDS

Series Editor: M. M. Koton, *Institute of Macromolecular Compounds Leningrad, USSR*

CAPILLARY LIQUID CHROMATOGRAPHY
B. G. Belen'kii, E. S. Gankina, and V. G. Mal'tsev

ION-EXCHANGE SORPTION AND PREPARATIVE CHROMATOGRAPHY OF BIOLOGICALLY ACTIVE MOLECULES
G. V. Samsonov

MECHANISMS OF IONIC POLYMERIZATION: Current Problems
B. L. Erusalimskii

POLYIMIDES: Thermally Stable Polymers
M. I. Bessonov, M. M. Koton, V. V. Kudryavtsev, and L. A. Laius

CAPILLARY LIQUID CHROMATOGRAPHY

B. G. Belen'kii
E. S. Gankina
and
V. G. Mal'tsev

Institute of Macromolecular Compounds
Academy of Sciences of the USSR
Leningrad, USSR

Translated from Russian by
R. N. Hainsworth

CONSULTANTS BUREAU • NEW YORK AND LONDON

Library of Congress Cataloging in Publication Data

Belen'kii, Boris Grigor'evich.
Capillary liquid chromatography.

(Macromolecules compounds)
Translation of:
Bibliography: p.
1. Capillary liquid chromatography. 2. Biopolymers—Analysis. 3. Macromolecules—Analysis. I. Gankina, E. S. II. Mal'tsev, V. G. III. Title. IV. Series.
QP519.9.C37B45 1987 543′0894 87-618
ISBN 0-306-10994-8

This translation is published under an agreement with the Copyright Agency of the USSR (VAAP).

A Division of Plenum Publishing Corporation
233 Spring Street, New York, N.Y. 10013

Printed in the United States of America

PREFACE

In the early 1980s capillary liquid chromatography was being established; it was a period in which only a few research groups published a relatively small number of papers on the subject. Interest has since taken off, and a period of intense development, to which no end is yet in sight, is now upon us. More investigators and instrument-making firms are now entering the field. This greater interest has resulted in the rapid appearance of two collections [1, 2] and a series of topical reviews [3-6].

However, it could hardly be said that all the problems in this area have been formulated, let alone solved. The preparation of very efficient – open tubular or packed – microbore columns, for example, remains more an art than a science, while the relationship between radial and longitudinal mass transfer, and the effect of transcolumn velocity profiles on chromatographic efficiency, have been very poorly studied. Indeed, recent publications on these subjects have sometimes, far from clarifying matters, only muddied them further.

Many instrument-making firms are trying to unify their equipment so that it is suitable for microbore, conventional (analytical), and preparative liquid chromatography. This approach has not realized the full potential of capillary chromatography, and there also remains room for improving the performance of capillary columns.

Therefore, when preparing this book, we have tried to consider only the most substantial questions. However, some aspects covered in [1, 2, 369] have been dealt with so well that we felt it pointless to give them much attention here. These aspects concern fast capillary liquid chromatography, the theory of superefficient capillary columns, the combination of capillary liquid chromatography and mass spectrometry, and postcolumn derivatization in capillary liquid chromatography.

On the other hand, many of the principal advances of capillary liquid chromatography, such as the development of the basic method [7-10], the creation of the refractive index microdetector [11], the fluorometric ultrasensitive analysis of multicomponent mixtures [12], and the analysis of the molecular-mass distributions of polymers [13], are contained in papers published in Russian and so are not widely known in the West.

We hope that this book will fill that gap.

Leningrad, August 1985

B. G. Belen'kii
É. S. Gankina
V. G. Mal'tsev

CONTENTS

Chapter 1

INTRODUCTION

Reducing the diameter of a liquid chromatograph column opens up a series of new perspectives. First, it decreases the expenditures of sorbent and eluent due to the smaller cross-sectional area (proportional to the square of the column diameter) of the column. Second, it lowers to the same degree the minimum detectable quantity of the test sample, for the same concentrational sensitivity of the detector, and this sharply increases the analytical sensitivity of the method. Microbore columns also have several more subtle advantages: an improvement in the radial heat transfer, and the consequent ability to increase the linear velocity of the mobile phase, which is otherwise limited by the heat generated by the friction of the liquid against the sorbent particles that in turn distorts the boundaries of the chromatographic zones. The better radial mass transfer and the smaller volumes of mobile and stationary phases make it easy and cheap to combine columns into systems 5-10 m long, thus having hundreds of thousands of theoretical plates. Small-diameter columns can be coupled "on line" with mass spectrometers or other efficient detectors, for which a large volume of effluent is undesirable.

It is also clear that a reduction in the volume and volumetric flow rate of the eluent must lead to liquid chromatography equipment becoming less expensive because of a decrease in the metal content of the pumps and the whole system of handling the solvent. A very important point is that a severalfold reduction in eluent volume significantly lowers the toxic and fire dangers of liquid chromatography, which is, in this respect, a hazardous technology. In turn, it then becomes possible to use expensive, specialized sorbents, or as eluents solvents that are either toxic or difficult to obtain.

In spite of its clear advantages, microbore liquid chromatography was not popular until recently, and the commercial availability of the requisite equipment was severely limited some years ago. It seems to us that there were three main reasons for this. First, there are certain technical difficulties in preparing microbore columns that are as efficient as conventional columns and in creating very sensitive detectors with volumes less than 1 μl, and pumps that can deliver eluent at 1-20 μl/min but with flow rate accuracies comparable with those of conventional pumps for modern liquid chromatography. Second, there seems to be a psychological barrier for analysts and researchers in acclimatizing to microtechnology for separation and analysis, especially for those accustomed to the scale of conventional chromatographic columns, with samples measured in tens or hundreds of microliters and retention volumes in milliliters or tens of milliliters. Third, perhaps the advantages of using microbore chromatography for dealing with many analytical problems were not understood, and this exacerbated the lack of interest instrument manufacturers showed in producing microbore equipment. After all, since they were successfully selling conventional chromatographic equipment that had been thoroughly tested, why start producing equipment that would compete with their basic products in what was a good market? It was also uneconomical for the consumers to change their operational stock of modern liquid chromatographs for equipment that makes it obsolete.

This is why, even now, the adoption of microbore columns for analysis is somewhat fraught with obstacles. Clearly, in order to expedite matters it is necessary first to overcome the technical difficulties involved in raising the precision of microbore liquid chromatography, or to define areas in which precision is not critical. Second, the design of microbore columns has to be worked over to make them comparable with conventional columns in terms of reliability, flexibility, and automation. Third, the advantages of microbore chromatography over the conventional sort must be clearly examined and the merits of the former must be exactly formulated.

It should be emphasized that microbore chromatography is not just a variant of modern liquid chromatography having a limited application. It will be seen later that this type of chromatography has set records in sensitivity of analysis, speed of separation, and efficiency of the chromatographic system.

We shall now look at some of the terminology. Liquid chromatography based on columns whose diameters (d_c) are less than 1 mm is called "microbore liquid chromatography," and this is the meaning we have used until now. The other term used for this field is "capillary liquid chromatography." The terms "ultramicrobore chromatography," using packed columns with diameters of 0.1-0.3 mm, and "semimicrobore chromatography," using columns 1.5-2 mm in diameter, are also accepted. (Note that here and henceforth the diameter referred to is, of course, the inside diameter. If otherwise, we say so explicitly.)

It seems to us that the most acceptable term, by analogy with gas chromatography, is capillary liquid chromatography. This common term would include microbore liquid chromatography, using packed microbore and ultramicrobore columns, and microcapillary liquid chromatography, which involves both open tubular and irregularly packed capillary columns.

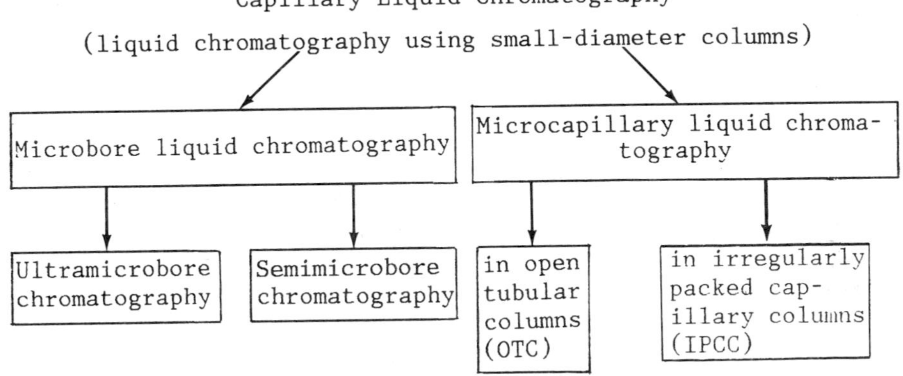

Since there is no real difference between the hydrodynamics of semimicrobore and ultramicrobore chromatography, the usefulness of further subdivision is in doubt.

The basic tasks of chromatography are to conduct analyses a) as rapidly as possible, b) getting the minimum peak volume so as to obtain the highest possible chromatographic peaks (this is in order to minimize the detectable sample mass or, equivalently, to maximize the analytical sensitivity), and c) as efficiently as possible so that the maximum number of components can be distinguished, i.e., maximize the peak capacity.

Recently, many experimental studies have been published on capillary liquid chromatography using packed columns, open tubular columns, and irregularly packed capillaries. Packed microbore columns are very successful, and those with diameters of 1 mm [14-16] and packed ultramicrobore columns with diameters of 0.25 mm [17-19] have been prepared that are as efficient as conventional columns with sorbent diameters of 5-10 μm. However, the experimental achievements using microcapillary columns are not as impressive. The point is that the efficiency of open tubular columns can be raised by reducing their diameters to the diameters of the sorbents used in modern efficient packed columns, i.e., to 5-10 μm. The best results so far have been achieved on columns 20-60 μm in diameter, which correspond to sorbents with approximately the same particle diameters. Reducing the diameters of open tubular columns to unit micrometer dimensions is hampered both by the difficulties in making such columns and by the absence of commercial devices compatible with them, i.e., injectors with subnanoliter volumes and solvent-delivery systems with flow rates around a nanoliter per minute. Even though open tubular microbore columns with diameters of 6 μm [20] and 10 μm [20, 21] have been obtained and tested, their properties fall far short of what is theoretically possible [22, 23].

As regards microcapillary columns, the originally suggested version [24-27] involved a clearly inefficient sorbent dimension, 30 μm, in a column 70 μm in diameter. These dimensions make severe demands on the volume of the measuring cell of the detector, which in this case does not confer to the system its necessary efficiency. However, the sorbent size for this type of column was later reduced to 10 μm [28], but the gain in efficiency proved smaller than expected. It seems we must therefore agree with Guiochon [29-31] that loosely packed microcapillary columns with 30-μm sorbents are a priori inefficient, and that packed capillary columns with 5-μm sorbents are not perceptibly preferable to open tubular columns 5-10 μm in diameter, which are simpler to construct. A common weak point of capillary columns (and mainly open tubular ones) is the fall in efficiency accompanying a rise in the solute's capacity factor.

An interesting, important topic concerns the smallest possible diameter for a microbore column. This dimension apparently depends on the type of pump and injector used. Assuming a reciprocating pump and rotary valve injector, as used for conventional liquid chromatography, then the minimum elution rate is 10-20 μl/min and the sample volume is 0.2-1 μl. The diameter of a packed microbore column is thus limited to 1-1.5 mm. If instead a special syringe micropump is used which can provide the requisite precision for deliveries of 0.1-10 μl/min, then it is possible to set up packed ultramicrobore columns with diameters from 0.2 to 0.5 mm. This raises the economy and sensitivity of the analysis five- to tenfold, compared to the 1- to 1.5-mm columns, but requires increased miniaturization of the measuring cell, the injector, and the other extra-column elements. Thus, the limiting diameter of the column in capillary chromatography depends on the technical capacities of the ancillary equipment and the need to use equipment common to conventional liquid chromatography.

The technical capacities of the ancillary equipment are also factors when it comes to choosing between open tubular and packed microbore columns for creating very efficient chromatographic systems. In order to achieve efficiencies of 10^6 theoretical plates a pressure greater than 300 MPa is necessary when using packed microbore columns (for a reasonable analysis time). Meanwhile, in order to get efficiencies of $5 \cdot 10^5$ plates using 5-µm open tubular columns under optimal conditions, the volume of the measuring cell must be significantly less than 1 nl. These limitations indicate that packed microbore columns should be used for efficiencies lower than $5 \cdot 10^5$ plates, while for efficiencies higher than 10^5-10^6 plates open tubular columns should be chosen.

Chapter 2

DEVELOPMENT AND ACHIEVEMENTS OF CAPILLARY LIQUID CHROMATOGRAPHY

The potential advantages of reducing the column diameter d_c to capillary dimensions were recognized by those investigating liquid chromatography long ago. The attempts by Stegemann and Bernhard [32] and Kirsten and Kirsten [33] to raise the analysis sensitivity to amino acids of classical systems involving cationite columns and a postcolumn ninhydrin reactor by reducing d_c to 0.5 mm deserve first mention. Horvath et al. [34], in 1967, were also investigating packed microbore columns (d_c = 1 mm) with a pellicular stationary phase. The first experiments with open tubular capillary liquid chromatographic columns (d_c = 0.23-0.3 mm) were carried out by Nota et al. in 1970 [35]. These early attempts were doomed to failure for two reasons, viz., the low efficiency of the columns and the absence of ancillary equipment suitable for capillary chromatography.

The history of capillary liquid chromatography really starts, therefore, with the papers of Sandakhchiev, Grachev, and Kuz'min [8-10], much of whose work remained unknown and was later duplicated elsewhere. In 1969 they described packed microbore columns with d_c of 0.5-1 mm which they employed to raise the sensitivity of chromatographic analyses of nucleic acids and their components. Their technique involved sample injectors 1-2 µl in volume, miniature syringe pumps that could deliver eluent at a flow rate of between 0.2 and 20 µl/min, and "on-column" spectrophotometric detection, i.e., the detector cell was the lower end of the column. Solvent-strength programming was achieved by creating a preformed composition gradient in the capillary before the analysis was started. This approach was developed by Ishii et al. [37-50], and we shall consider their work in more detail in Chapter 4.

Sandakhchiev et al. demonstrated that not only could columns with diameters of 0.5-1 mm separate multicomponent mixtures as well as conventional columns, but the capillary columns had lower sample dilutions in the column; hence the sample detectabilities were higher. They showed [7-10, 36] that ribonucleotides, the ribonuclease hydrolyzates of RNA, oligonucleotides, and mono-, di-, and triphosphonucleotides could all be separated using microbore columns packed with ion-exchange cellulose, Dowex 50 × 4 cationite, or (using steric-exclusion chromatography) Sephadexes. Since these pioneers of capillary liquid chromatography used soft or semirigid large-particle swelling sorbents, they left a series of problems unresolved. Essential for obtaining a high efficiency in liquid chromatography, these problems were related to the use of rigid microparticle sorbents and high pressures.

The high-performance variant of microbore chromatography was developed by Ishii et al. [37-50]. Their equipment was designed to operate under pressures of 10 MPa and used spectrophotometer detection in a quartz capillary to minimize the extracolumn spreading. This system was incorporated by JASCO in the first commercial capillary liquid chromatograph, the Familic 100 (Fast Microbore Liquid Chromatograph).

Ishii et al. made their columns from Teflon tubing 0.25-0.5 mm in diameter cut to lengths of 3-20 cm. The stationary phase was packed into the column under pressures greater than the chromatograph's working pressure and was thus subject to a radial compression by the elastic walls of the Teflon column. They believed that the wall effect (see Section 5.2.1) in columns packed with pellicular sorbents (d_p = 30 μm) or microparticle sorbents (d_p = 5 μm) is reduced. Ishii et al. have described their methods for sealing the Teflon columns, packing in the sorbent particles, stop-flow sample injection, and elution using a gradient preformed in a storage capillary with a double-syringe pump. They have also developed a method for concentrating the samples using subsidiary columns (precolumns) in front of the analytical system.

Ishii's group have successfully applied microbore columns to a variety of chemicals and have used every sort of liquid chromatography, i.e., adsorption, reversed-phase (hydrophobic), ion-exchange, and steric-exclusion chromatographies. Unfortunately, the Teflon columns they have based their work on are not particularly efficient. The shortest HETPs they have achieved are about 0.4 mm for surface-porous sorbents (d_p = 30 μm) and 36 μm for microparticle sorbents (d_p = 5 μm). This is far from the efficiencies that would correspond to the best obtainable in conventional high-performance packed columns, for which $H_{min} = 2d_p$.

In order to improve the situation, Ishii's group then developed ultramicrobore packed columns 0.1-0.2 mm in diameter [18, 45-50].

When the columns were made from Teflon tubing, an ultramicrobore column 0.15 mm in diameter had an efficiency a factor of 1.5-3 times worse than a column 0.5 mm in diameter. However, making the column from other materials significantly increased the efficiency. Thus, for ultrathin Pyrex columns $H_{min} = 5d_p$ [45], while for fused-silica capillaries H_{min} is between $2.5d_p$ and $3d_p$ [18, 46, 47]. By connecting together a series of seven short (20 cm long) ultramicrobore columns made from fused silica ($d_c = 0.35$ mm), Takeuchi and Ishii [46] were able to obtain a steric-exclusion chromatography regime that had an overall efficiency of 100,000 theoretical plates and an analysis time of 80 min. Yang [17] continued the work on packed fused-silica columns using a sorbent with $d_p = 3$ μm in systems up to 2 m long. At the optimal mobile phase linear velocity of 0.15 cm/sec the minimum HETP was $3.2d_p$, yielding an overall efficiency of 110,000 theoretical plates for a column 1 m long.

Fused-silica tubing has, in fact, a number of merits: 1) It can be used with pressures up to 80 MPa. 2) Its internal surfaces are smooth, which means that a column up to 1 m long can be packed without there being any significant contribution to the band spreading from the wall effect. 3) Its optical transparency means that the column itself can function as the measuring cell for an on-column UV spectrophotometer or fluorometer. 4) The small outside diameters allow the tubes to be connected directly to the injectors or detectors, thus eliminating additional extracolumn volumes. And, finally, 5) the columns can be easily coiled into spirals because of the flexibility of fused-silica capillaries. Ultramicrobore silica capillaries have been successfully used for high-speed microcapillary liquid chromatography [48], to analyze trace elements [49], and to separate complicated multicomponent mixtures both with gradient regimes [50] and with temperature programming [51].

The story continues with the work of Scott and Kucera and others [14-16, 52-54] on high-efficiency and superfast chromatographies at high pressures. They employed packed steel microbore columns 1 mm in diameter, which were packed at a pressure of 175 MPa with a stationary phase with d_p of 5-20 μm. Until Scott and Kucera published their papers the mainstream of work had been striving for the realization of the analysis sensitivity and solvent-economy promise of capillary chromatography. However, the papers demonstrated that there were two further potential advantages, i.e., superefficient systems made up of several connected columns and with N values of up to a million theoretical plates [16, 52], and superfast analyses (with productivities of 280 theoretical plates/sec [54] that could be attained with velocities of 12 cm/sec).

In conventional liquid chromatography there are only two ways of getting very high efficiencies: either the effluent is recycled, or short, efficiently packed columns are joined together (H_{min} =

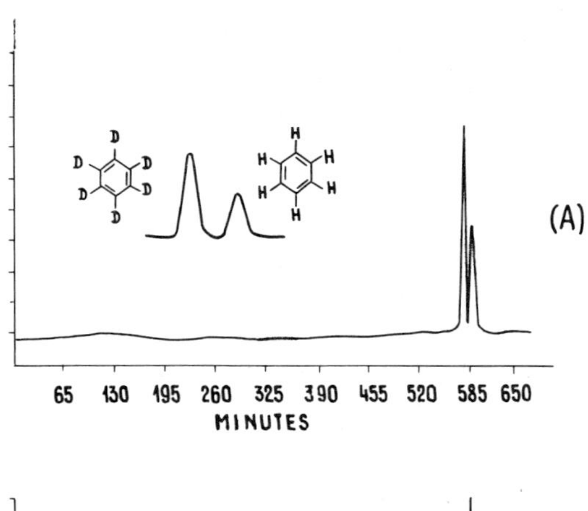

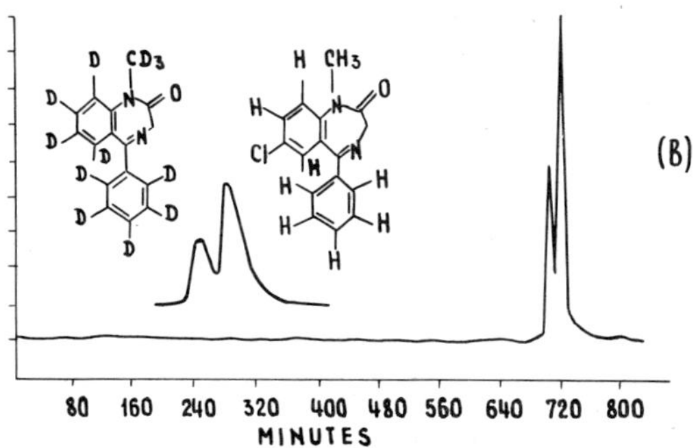

Fig. 2.1. Separating mixtures of benzene and benzene-d_6 (A) and diazepam and diazepam-d_{11} (B) using a 4.5 m × 1 mm column filled with ODS silica gel (N = 230,000 theoretical plates). Eluent: an 85:15 mixture of methanol and water, 10 μl/min. Sample volume: 0.5 μl; detector: UV sensor for 254 nm. (Reproduced from [16] with the permission of Elsevier Science Publishers.)

$2d_p$). The number of separation recycles is restricted by the admissible spreading of the samples. On the other hand, Kucera and Manius [16] have found that when conventional columns are connected together there is a 60% loss in efficiency at each stage. Even though this conclusion contradicts results obtained by other workers [55, 56], the value of microbore columns for creating superefficient systems is indubitable, if only for their economy of the mobile and stationary phases. That tens of microbore columns 50 cm long can be connected in series while retaining 100% efficiency has been demonstrated by Kucera and Manius [16] and Scott and Kucera [52]. They put together a system that had an overall efficiency of $7 \cdot 10^5$ theoretical plates and which could separate substances having only minor structural differences, such as the hydrogen and deuterium forms of benzene and diazepam (Fig. 2.1).

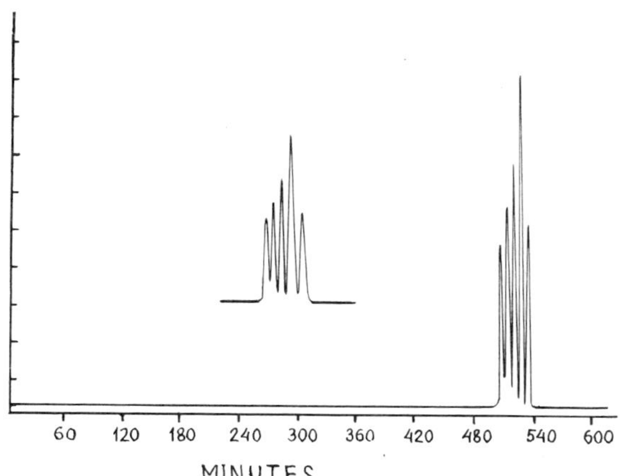

Fig. 2.2. Separation of alkylbenzenes (C_2-C_6) using steric-exclusion chromatography in a 14 m × 1 mm column filled with Spherisorb (d_p = 5 µm), N = 650,000 theoretical plates. Eluent: tetrahydrofuran, 25 µl/min. Sample: 0.5 µl of 10% C_2-C_6 alkylbenzenes; detector: UV sensor for 254 nm. (Reproduced from [16] with the permission of Elsevier Science Publishers.)

A good application of superefficient capillary columns is the steric-exclusion chromatography of low-molecular-weight compounds and oligomers. Because the range of possible distribution coefficients is narrow ($0 \leqq k_d \leqq 1$), this method has a relatively low selectivity, which is aggravated for low-molecular-weight compounds by the absence of "rigid" silica-based sorbents that have a sufficient specific pore volume (V_p = 1 ml/g) but small pore diameters ($D_p \leqq 50$ Å). This low selectivity can nevertheless be compensated by the large efficiencies, as shown in Fig. 2.2 for the steric-exclusion capillary chromatography of alkyl-substituted benzenes.

The papers by Scott, Kucera, and Munroe [52, 54] indicated that superfast analyses could be much more easily implemented on microbore columns than on conventional columns.

The frictional heat generated by the liquids as they pass at very high velocities through the pores cannot be dissipated by large-diameter columns; moreover, the radial temperature gradient that is set up adds to the zone spreading [57]. We can expect this effect to have less influence in microbore columns because of their good radial heat-exchange properties. Although this aspect remains unproven because no one has made any quantitative comparisons, rapid capillary chromatographs have the certain advantages of slow volumetric flow rates, i.e., eluent economy and longer pump service

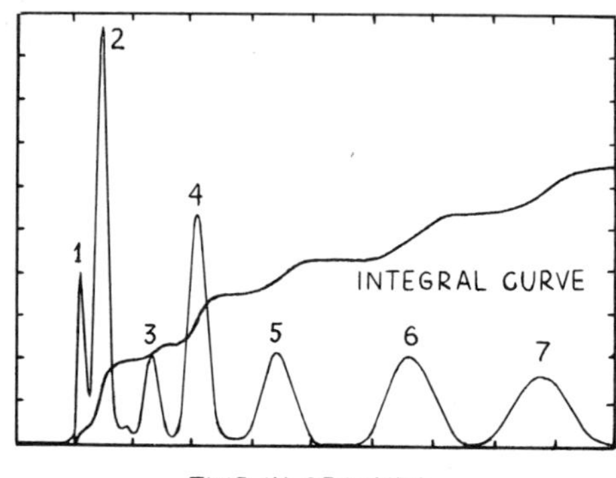

Fig. 2.3. Rapid microbore chromatography of diazepam and its metabolites on a 1500 mm × 1 mm column packed with Partisil-10 (d_p = 10 μm). Eluent: methanol—ethyl acetate—heptane (8:10:82), 1.5 ml/min; N/t_R = 70 theoretical plates; UV detection at 254 nm. (Reproduced from [54] with the permission of Elsevier Science Publishers.)

lives. An example of a rapid analysis is the 30-sec separation of a seven-component mixture [54] which consumed a total of 2.2 ml of eluting solvent (Fig. 2.3). High-speed chromatographic analyses open up possibilities for enhanced therapeutic monitoring of medicines [58, 59] and for studying the kinetics of fast reactions [60] because it will be possible to measure the changes in the reaction compositions at short intervals of time.

Scott et al. have also developed an efficient way of packing stable stainless-steel microbore columns, which made it possible to obtain columns 100 cm long and d_p of 20 μm [52] or 50 cm long and d_p of 8 μm [14, 16] and having reduced HETPs of $h_{min} = H_{min}/d_p = 2$. This value is characteristic of the best that researchers have achieved on conventional high-performance columns. They have also made significant contributions to the design of ancillary apparatus for microbore columns based on equipment available for high-performance conventional liquid chromatography, thus raising the pressures that can be routinely supported in microbore columns to 50 MPa, enabling the gradient to be controlled automatically [53], and enhancing other equipment [52, 54]. This work has been incorporated into modern chromatographs such as the Familic 300 (JASCO) or the LC-5A (Shimadzu), which are described in Chapter 4. The result of Scott and Kucera's work, and its development by others [61-67], is that stainless steel microbore columns (d_c = 1 mm) are now the standard analytical tool for liquid chromatography.

Side by side with the development of microbore liquid chromatography in the late 1970s and by analogy with capillary gas chromatography went the spread of capillary liquid chromatography using open tubular or irregularly packed capillary columns. The stimulus for the work was not so much to increase the sensitivity, as to increase the speed of superefficient analyses and to realize the potential advantages of slow flow rates. There was even no expectation that there would be large gains in sensitivity because to achieve them the detectors would have to have measuring cells 1-10 nl in volume [22, 23], which for concentration-sensitive detectors would lead to such a loss in detection sensitivity that the advantages of reducing the column diameter would be mostly cancelled out. The main reason for using capillary columns was to increase efficiency because it is very difficult to pack very long conventional columns well, or to connect them together. Moreover, the lower permeabilities of regularly packed columns restricted their lengths.

The typical flow rates for capillary columns is 0.1-1 μl/min, which, as has been pointed out [21, 68-72], is a major advantage, for it makes it possible to employ types of detectors that had previously been considered unsuitable for high-performance liquid chromatography, such as mass spectrometers [21, 68, 69], and the flame ionization [70, 71] and flame emission [71, 72] detectors that have been developed for gas chromatographs.

The application of open tubular capillary columns is associated with the work of Tijssen et al. [21, 73], Hibi, Tsuda, Takeuchi, Ishii and others [74-90], and Krejči et al. [70, 91-93], who developed the methods for preparing (see Chapter 5) and coating them with stationary phase, and who described appropriate sample introduction, solvent delivery, and solute detection techniques. The abilities of open tubular capillary columns, with diameters of 0.03-0.1 mm, have been demonstrated for adsorption [79, 80, 82], distributive [76, 82], reversed-phase [75, 85], and ion-exchange [81, 90] chromatographies. Tsuda, for instance, has obtained [87] an efficiency of $1.6 \cdot 10^5$ theoretical plates (4620 theoretical plates/m) for an open tubular capillary column 28.5 μm in diameter and 35 m long, coated with β,β'-oxydipropionitrile (ODPN). Subsequently, Krejči et al. [91] demonstrated that an open tubular capillary column could, in principle, be prepared with a theoretical plate count of $1.25 \cdot 10^6$, and that open tubular capillary columns could be applied to an ultrasensitive analysis (around 0.05 pg) by using an electrochemical detector with a high concentrational sensitivity and a volume of 1 nl [93]. Tijssen et al. [21] investigated open tubular capillaries 10-50 μm in diameter and coated with ODPN or Apiezon L. They demonstrated that extremely complicated mixtures could be separated with efficiencies of up to $5 \cdot 10^6$ theoretical plates.

The next important step was the formulation by Knox [22, 23] of a theory for the optimal regimes for capillary liquid chromatography and all its variants. He showed that a consequence of the greater permeability of open tubular capillary columns is that they will yield faster analyses than packed columns if very high efficiencies are required. In order for open tubular capillaries to retain this advantage for $N \approx 10^5$ theoretical plates their diameters have to be reduced to 1-10 μm, i.e., to the dimensions of a sorbent grain in a packed high-performance column. Moreover, the extracolumn volume has to be lowered to well below 1 nl, which is technically very difficult. How capillary liquid chromatography can be optimized is considered in detail in Chapter 3.

Considerable efforts have been recently exerted to realize the potentially very high efficiencies of open tubular capillary columns ($d_c \leqq 10$ μm). The successes that have been achieved in the creation of very efficient reversed-phase open tubular capillaries (d_c = 30-40 μm) [85, 94] and their application to the gradient chromatography of multicomponent mixtures [85] must be noted. And now open tubular columns with diameters of 6 and 10 μm [20] have been prepared and tested, with good results, the efficiencies reaching $5 \cdot 10^4$ theoretical plates for a column 830 cm long and a ΔP of 19 MPa, i.e., 50-60 theoretical plates/sec. The chromatogram in Fig. 2.4 is for an eight-component mixture that was separated in 6.5 min on an open tubular capillary column (d_c = 6 μm).

Other interesting attempts include the use of planar field-flow fractionation channels for high-performance capillary columns [95] or single-phase (or hydrodynamic) [96] capillary liquid chromatography in open tubular columns. The efforts to get around the stringent requirements on the size of the detector cells and injectors and to increase the capacity of open tubular capillary cells have resulted in a number of investigators [97] looking at the idea of parallel flows in open tubular capillaries. Unfortunately, this approach does not seem practicable because of the difficulties of ensuring identical flow velocities in each of the capillaries, as the first experiments have demonstrated [97].

A third type of column has been developed by Tsuda and Novotny [24-28]. This is the irregularly packed capillary column 70 μm in diameter, for which $d_c/d_p < 5$. The merit of this type of column is that it has a relatively high permeability. The stationary phase is formed in the capillary by sintering Al_2O_3 particles or silica gel (d_p = 30 μm) in it as it is formed. This sort of column can also be prepared by taking a conventional, regularly packed glass column with a suitable diameter (200-300 μm) and drawing it out at a high temperature to the required length. This method can yield a packed capillary column up to 40 m long and with an efficiency at moderate pressures (9.8 MPa) of 85,000 theoretical plates. The stationary phase of a packed capillary column can also be modified, as described in [26], to produce sorbents suitable for re-

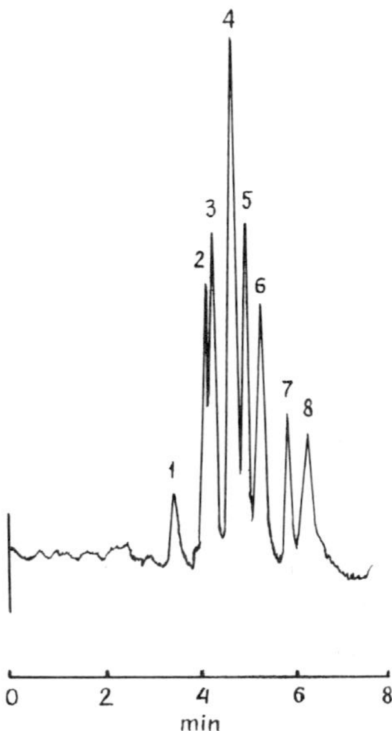

Fig. 2.4. Chromatogram of isomers of chloroaniline and chlorotoluidine on a 2200 mm × 6 μm open tubular capillary column coated with a quasi-silica gel. Mobile phase: n-heptane containing 0.7% acetonitrile, 0.7% methanol, 0.3% dichloromethane, and 0.01% water. The sample contained 1) o-, 7) m-, and 8) p-chloroaniline; 2) 3-chloro-p-, 4) 4-chloro-o-, and 6-chloro-o-toluidine; and 3) o-, 5) m-, and 6) p-toluidine. Pressure 11 MPa; UV detection at 235 nm. (Reproduced from [20] with the permission of Elsevier Science Publishers.)

versed-phase, normal-phase, and ion-exchange chromatographies. Unfortunately, the HETP for these columns begins to rise quite sharply if the elution velocity is raised above 0.7 cm/sec. This unusual behavior is associated [24, 25] with the irregularity of the packing, but it has not yet been convincingly explained theoretically. This prevents packed capillary columns from competing with open tubular ones in terms of speed of analysis and efficiency.

If sorbent particles smaller than 30 μm in diameter are used, then the stability of an irregularly packed column, and hence its efficiency, falls. These serious flaws in packed capillaries led to Guiochon's criticisms [29, 30], which were so convincing that recently even the initiators of this research have begun to use fused-silica packed ultramicrobore columns with regular packing [19]. Nevertheless, the development of packed capillary liquid chromatography, together with the efforts [20, 21, 70] to prepare the open

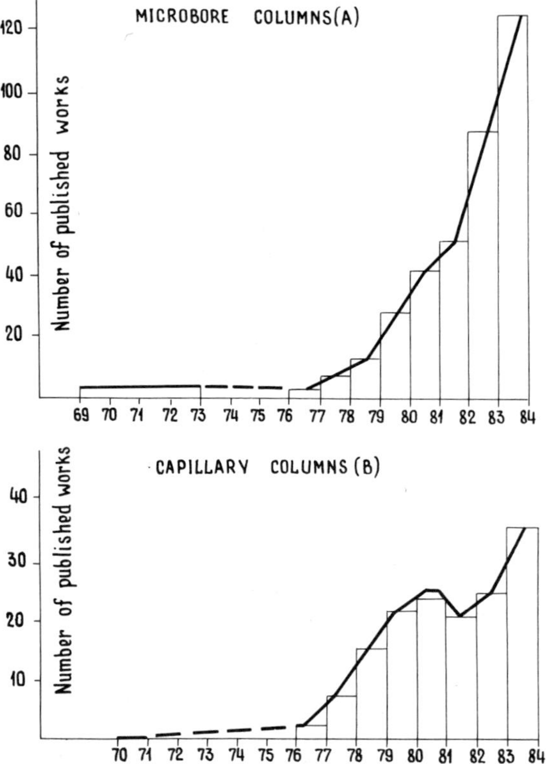

Fig. 2.5. Growth in the annual publication rate on microbore liquid chromatography A) in packed columns and B) in capillary columns, both open tubular and with irregular packing.

tubular capillary columns with diameters less than 10 μm that were recommended by Knox's theory, has stimulated work on improving the ancillary equipment. The more important achievements in this respect have been on-column photometric [98, 99], fluorometric [100, 101], electrochemical [102], and potentiometric [103, 104] detectors with cell volumes less than 1 nl; devices for delivering eluents with gradients at low speeds [105, 107] and for injecting nanoliter samples [108]; temperature programming [51, 62, 64]; and the connection of columns to mass spectrometers [68, 109-123] or IR spectrometers with Fourier transform capacities [124-128].

An important advance for ultrasensitive capillary liquid chromatography was the development of sensitive detectors with cell volumes smaller than 1 μl. The following are good examples:

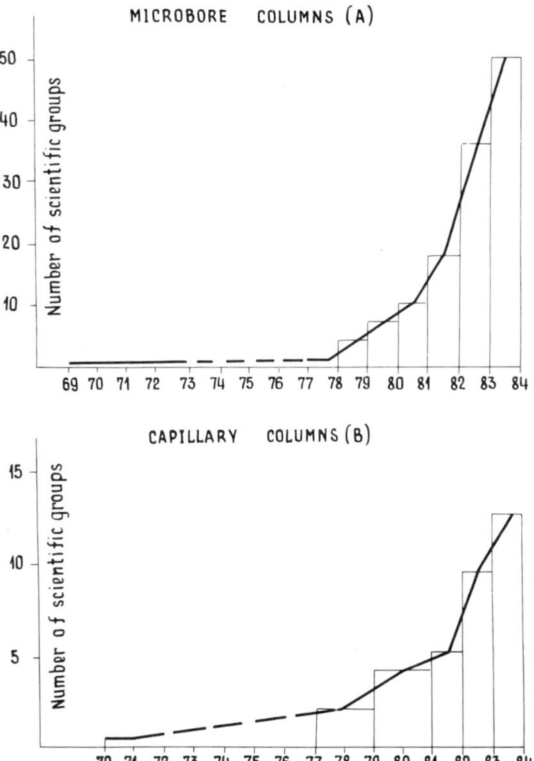

Fig. 2.6. Annual increase in number of research groups publishing papers on microbore columns A) for packed columns and B) for capillary columns.

the μLC-10 spectrophotometer detector (ISCO), which has a cell volume of 0.5 μl and an optical path of 1 cm; laser fluorometers [101, 129-131]; and electrochemical detectors [93, 102], particularly the parallel opposed double-electrode type, which allows more signal amplification the smaller the dimensions of the cell [132]. This topic is considered in more detail in Chapter 4.

The latest step is the development of the methodological principles for analyzing multicomponent mixtures on capillary liquid chromatographs. It should be remembered that an immense distance stretches from the simple demonstration that a method is possible to the application of the method for quantitative analyses. For a mixture with a complicated composition to be analyzed quantitatively it is essential that the sample injection, retention volumes, and peak heights for gradient elution be reproducible, and that the calibration curves in steric-exclusion chromatography be linear for a wide range of molecular masses. These problems have been addressed in a series of papers [12, 13, 53, 133-135].

An interesting and very promising area of analytical chemistry, which has grown up in parallel with capillary liquid chromatography, is supercritical-fluid chromatography [136-143]. The mobile phase in this technique is a compound that is kept at a temperature and pressure above its critical point; thus the phase has properties intermediate between those of a gas and those of a liquid. Microbore columns, because they can be rapidly heated and because of their good radial heat-exchange properties, are particularly suited to this sort of chromatography (see Section 5.4.1).

Microcapillary chromatography is now being developed at a furious pace, which can be seen by the avalanche of papers and ever-growing number of research groups who are developing and using this method. The growth of the field can be seen from Figs. 2.5 and 2.6. The interest in packed microbore columns has grown more quickly than the interest in capillary columns. The period from 1980 to 1983 was characterized (Fig. 2.5B) by a relatively stable annual level of publications on capillary columns, clearly the result of the theoretical criticisms leveled by Knox [22, 23] and Guiochon [29, 30]. In 1983-1984 the interest in capillary columns revived and efforts were made to overcome the volumetric limitations and the problems of dispensing and detecting subnanoliter volumes.

Chapter 3

THEORETICAL BASIS OF CAPILLARY LIQUID CHROMATOGRAPHY

3.1. BAND SPREADING IN PACKED AND OPEN TUBULAR COLUMNS

Band spreading in a chromatographic column is related to the stochastically different paths taken by individual molecules (particles) as they are carried by the mobile phase. Their overall retention times differ because of Brownian motion (molecular diffusion), the different residence times in the channels of the packing, in which the mobile phase moves at different rates (velocity profile effects), the different numbers of contacts with the sorbent surfaces (interparticle diffusional mass transfer), the different residence times in the stagnant zones of the mobile phase (mass transfer in the stagnant zones), and the different residence times in the stationary phase (intraparticle diffusional mass transfer). The contributions of these processes to the total peak variance are independent and therefore additive, viz.,

$$\sigma^2_{L,c} = \sum_j \sigma^2_{L,i}, \tag{3.1}$$

where $\sigma^2_{L,c}$ is the column variance in units of length, and $\sigma^2_{L,i}$ are the individual process variances (length units).

Instead of the total variance itself, we can use its increments per unit length of the column $d(\sigma^2_{L,i})/dL$, i.e., the height equivalent to a theoretical plate (HETP) $H_i = d(\sigma^2_{L,i})/dL$,

$$H = \sum_j H_i. \tag{3.2}$$

TABLE 3.1. Contributions of Different Processes to the HETP [144]

No.	Process	Contribution to HETP*	
1	Spreading due to flow profile	$C_\ell d_p$	$\bar{A}u^{1/3}$
2	Mass transfer in mobile phase	$C_m d_p^2 u/D_m$	
3	Longitudinal diffusion	$C_d D_m/u = \bar{B}/u$	
4	Mass transfer in stagnant mobile phase (in sorbent pores and between particles)	$C_{s,m} d_p^2 u/D_m = \bar{C}u$	
5	Mass transfer in stationary phase	$C_s \cdot d_f^2 u/D_s = \bar{C}'u$	

*C_ℓ, C_m, C_d, $C_{s,m}$, and C_s are coefficients for a well-packed column; d_p is the diameter of a sorbent grain; d_f is the film thickness of the stationary phase; u is the linear velocity of the mobile phase; D_m is the diffusivity in the mobile phase; D_s is the diffusivity in the stationary phase; and $\bar{A}$, $\bar{B}$, $\bar{C}$, and $\bar{C}'$ are constants for a given column.

How H and its individual components H_i depend on the chromatographic parameters, primarily the linear velocity of eluent, and optimizing the method with respect to the column parameters are the topics of many investigations [144-149]. Our current understanding of the contributions the sources of spreading make to the overall HETP for packed columns is laid out in Table 3.1.

In accordance with Table 3.1, a common form for the equation for the HETP is

$$H = \bar{A}u^{1/3} + \frac{\bar{B}}{u} + \bar{C}u + \bar{C}'u. \tag{3.3}$$

Instead of H and u we shall use the dimensionless numbers $h = H/d_p$ (the reduced plate height) and $\nu = ud_p/D_m$ (the reduced velocity), which Giddings introduced [144]. These enable us to obtain a very general expression that is free from both the diameter of the sorbent beads and the diffusion coefficient of the substance being analyzed, i.e.,

$$h = (\bar{A}D_m^{1/3}/d_p^{4/3})\,\nu^{1/3} + (\bar{B}/D_m)1/\nu + [(\bar{C} + \bar{C}')D_m/d_p^2]\nu. \tag{3.4}$$

Or by substituting the bracketed expressions by constants

$$h = \frac{B}{\nu} + A\nu^{1/3} + C\nu. \qquad (3.5)$$

This equation was given by Knox [145, 147], and the first term is the contribution to the HETP of molecular diffusion, the second term the combined contribution of eddy diffusion, convection, and velocity profile, and the third term is the contribution of mass transfer between the stationary and mobile phases. For tolerable columns we have $A < 2$, $B < 4$, and $C < 0.2$, while for well-packed columns $A \approx 0.5$, $B \approx 2$, and $C \approx 0.05$ [150]. A well-packed column has a small h_{min} of around 2. If the mass transfer is good, then the slope of the h versus ν curve will be small for large ν. For pellicular sorbents like Zipax, C is less than 0.02, while for tightly linked ion-exchange resins C is around unity [150]. If the column is well packed, then according to [22, 23] we can, for $k' = 2$, rewrite (3.5) as

$$h = \frac{1.8}{\nu} + 1.7\nu^{1/3} + 0.05\nu. \qquad (3.6)$$

Other values for the coefficients in (3.5) are given in the literature for even more efficient packed columns [30, 58], i.e.,

$$h = \frac{2}{\nu} + \nu^{1/3} + 0.05\nu. \qquad (3.7)$$

In contrast to packed column chromatography there is an exact theory for open tubular chromatography, which was developed by Golay in 1958 for gas chromatography [151]. Using the reduced variables Golay's equation becomes

$$h = \frac{2}{\nu} + \frac{1 + 6k' + 11k'^2}{96(1 + k')^2}\nu + \frac{2}{3}\,\frac{k'}{(1 + k')^2}\left(\frac{d_f^2}{d_c}\right)\left(\frac{D_m}{D_s}\right)\nu. \qquad (3.8)$$

Here k' is a capacity factor and is the product of the distribution coefficient k_d and the ratio of the volumes of the stationary and mobile phases V_s/V_m, D_s is the diffusion coefficient in the stationary phase, and d_f is the film thickness of the stationary phase.

The last term in (3.8) is relatively small for gas (and more so for liquid) chromatography, because it is related to the mass transfer in a thin film of the stationary phase. Hence it is possible in practically every case to consider that

$$h = \frac{2}{\nu} + \frac{1 + 6k' + 11k'^2}{96(1 + k')^2}\nu = \frac{2}{\nu} + 0.066\nu, \qquad (3.9)$$

the final expression in (3.9) being the result for $k' = 2$. It is easily seen that for $k' = 2$

$$h_{min} \equiv h_0 = \frac{1}{1 + k'}\left(\frac{1 + 6k' + 11k'^2}{12}\right)^{1/2} = 0.72, \qquad (3.10)$$

which corresponds to $\nu_{opt} \equiv \nu_0$:

$$\nu_{opt} \equiv \nu_0 = (1 + k')\left[\frac{192}{(1 + 6k' + 11k'^2)}\right]^{1/2} = 5.1. \qquad (3.11)$$

Other values of h_0 and ν_0 correspond to other values of k' (see Fig. 3.1).

As for packed columns, according to Eq. (3.6) we get

$$h_0 = 3.16 \text{ for } \nu_0 = 2.7 \text{ and } k' = 2, \qquad (3.12)$$

although Eq. (3.7) gives different values, i.e.,

$$h_0 = 2.15 \text{ for } \nu_0 = 3 \text{ and } k' = 2.$$

These are the values usually used for calculating the efficiency of packed columns. In our work on optimization we shall use the coefficients in (3.6) because they are more applicable to microbore chromatography. Clearly, these results must agree qualitatively with those obtained using (3.7).

For an unretained solute (k' = 0) Golay's equation turns into Taylor's equation [152]

$$h = 2/\nu = \nu/96. \qquad (3.13)$$

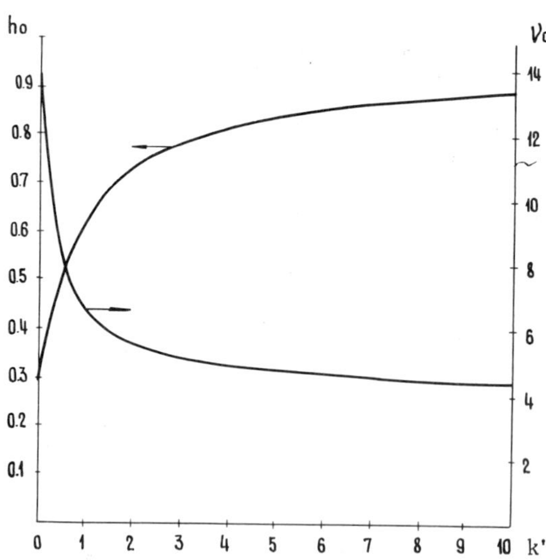

Fig. 3.1. Minimum reduced HETP and optimum reduced velocity versus capacity factor for open tubular columns [as per Eq. (3.9)].

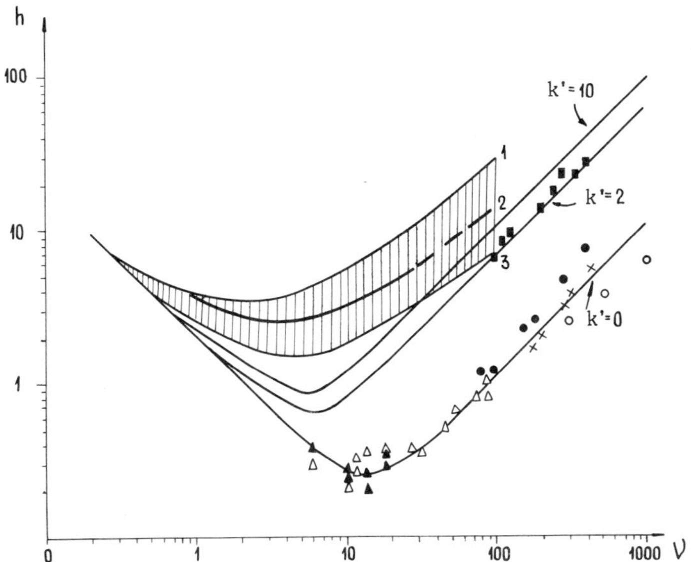

Fig. 3.2. Reduced HETP versus reduced velocity for packed microbore columns (curves 1-3) and an open tubular column. Curve 1 is from Eq. (3.6), curve 3 from Eq. (3.7), and curve 2 from Kucera's experimental data [16]. The open tubular points are taken from [104, 75, 310].

This describes the spreading of a substance zone as it moves through an empty capillary (without the stationary phase).

The way h depends on ν for packed columns and open tubular columns is shown in Fig. 3.2. The values of h for packed columns are scattered within a certain range (shaded in the figure) because the constants in (3.5) depend on the quality of the packing. Equation (3.9) is exact for open tubular columns and describes the experimental data very well, as can be seen from Fig. 3.2.

3.2. OPTIMIZING LIQUID CHROMATOGRAPHY

The basic task of chromatography is the separation of substances, and this is overwhelmingly dependent on the quality of the main chromatographic instrument, the separating column. This quality can be thought of as having three elements: a) selectivity, which is the ability to separate substances that differ in their chemical structures, b) efficiency, which is the ability at a given eluent flow rate to generate narrow peaks that can be used to obtain a separation when the selectivity is limited, and c) the hydraulic permeability, which regulates the pressure required to achieve an eluent flow rate that will, for a given selectivity and efficiency, yield a desirable analysis time.

The separating power of a column for a two-component mixture is described by the resolution K_R. This is the ratio of the difference between the retention volumes of two peaks ($\Delta V_R = V_{R,2} - V_{R,1}$) to double the sum of their volumetric standard deviations ($\sigma_{V,1}$ and $\sigma_{V,2}$):

$$K_R = \frac{1}{2}\frac{V_{R,2} - V_{R,1}}{\sigma_{V,1} + \sigma_{V,2}} = \frac{1}{4}\frac{\Delta V_R}{V_R}N^{\frac{1}{2}} = \frac{1}{4}\frac{\Delta k'}{1 + k'}N^{\frac{1}{2}} =$$

$$\frac{1}{4}(2 - 1)\frac{k'}{1 + k'}N^{\frac{1}{2}}, \qquad (3.14)$$

where N is the efficiency of the column in theoretical plates, and $\alpha = k_2'/k_1'$ is a selectivity parameter. Equation (3.14) also contains the quantities V_R, N, and k' averaged over both peaks.

It can be seen that the resolution is the product of three components, which are related to the selectivity α, the capacity factor k', and the efficiency N of the column. In order to obtain the necessary resolution, any of these components may be varied. However, when a two-component mixture is to be separated, it is useful to raise the selectivity $\alpha = k_2'/k_1'$ and reduce k' so that the product of the first and second components in (3.14) is increased. This way the efficiency N of the column can be decreased. This makes it possible to contract both the separation time and the retention volume, which are both proportional to (1 + k') and N.

Note that the above method of optimizing the resolution's components can be clearly seen from the third of the expressions in (3.14), viz.,

$$K_R = (1/4)[\Delta k'/(1 + k')]N^{\frac{1}{2}} = (1/4)SN^{\frac{1}{2}}.$$

This was used in [149] for optimizing the liquid chromatography of polymers.

When a mixture of several substances is to be separated, another criterion is introduced. This is the peak capacity n_p [144], which is the number of peaks in a chromatogram that can be resolved for a resolution of unity (though other values of K_R can be taken), i.e.,

$$n_p = 1 + 0.6N^{\frac{1}{2}} \log (1 + k_n'), \qquad (3.15)$$

where k_n' is the capacity factor for the last peak. For $K_R = 1$ it is convenient to determine the peak capacity for $k_n' = 6.4$, whence $n_p \approx 0.5N^{\frac{1}{2}}$.

It is clear that the selectivity α of the separation of each pair of peaks must be such as to ensure the complete resolution of the maximum number of peaks in the chromatogram (for a given K_R, N, and k_n'). To achieve this it is necessary to use gradient elution or to program other chromatographic parameters.

There is one sort of chromatography in which the efficiency is determined by the column selectivity other than via (3.14). This is the steric-exclusion chromatography of polymers. Belen'kii and Vilenchik have shown [149] that the column selectivity, which in this case is given by

$$V_R = C_1 - C_2 \log M, \tag{3.16}$$

is related to N thus:

$$N \geqq 7.3/(C_1/C_2 - \log M). \tag{3.17}$$

It can be seen from (3.17) that when the molecular mass M of the largest polymer homolog is raised the column efficiency must rise.

Hence (3.14), (3.15), and (3.17) can be used to determine, for a given selectivity, the column efficiency needed to obtain the required resolution when analyzing mixtures with two or more components or polymer homologs.

The problem of the selectivity of a chromatographic experiment is very complicated and in most cases ambiguous, in that the selectivity depends not only on the substances being separated, but also on the abilities of the experimenter himself. We shall not discuss this problem further. We shall consider that we have a chromatographic system of the requisite selectivity.

We now want to look more closely at the performance of the system. We shall refrain for a while from defining performance and only indicate that at least three of the tasks involved in chromatographically separating solutes are related to it. They are the problems of: maximum speed; minimum chromatographic zone width, which corresponds to the maximum analytical sensitivity or the purity of the collected fractions; and maximum efficiency, which determines the peak capacity of the column, i.e., the number of separable components.

As we mentioned above, three types of separating columns are presently used for liquid chromatography. They are:

1) packed columns (PC) and packed microbore columns (PMC), which are compactly filled with sorbent particles that are either spherical or irregularly shaped, and for which the ratio of the column diameter to the sorbent grain diameter is $d_c/d_p \gg 10$;

2) open tubular columns (OTC), on whose internal surface there is a thin silica-gel film onto which the organic stationary phase may be grafted; and

3) packed capillary columns (PCC), the diameters of whose sorbent beads are $d_p \cong 30\ \mu m$ and for which $d_c/d_p \cong 3$.

The three column types have different hydrodynamic permeabilities k_0, which can be determined from Darcy's equation [153]

$$u = k_0 d^2 \Delta P / L\eta, \tag{3.18}$$

or by using the reduced HETP, h, and the reduced linear velocity ν:

$$\nu = k_0 d^2 \Delta P / Nh\eta D_m, \tag{3.19}$$

where u is the linear eluent velocity, $d = d_p$ is the mean diameter of a sorbent grain for packed columns or $d = d_c$ the diameter of the capillary for open tubular columns, and L is the length of the column.

For packed columns with regular packing ($d_c/d_p \gg 10$), k_0 ranges from 1/400 for impermeable spheres to 1/1550 for porous, irregularly shaped particles [153]. In practical calculations an intermediate value can be taken, i.e., $k_0 = 10^{-3}$, while for open tubular columns $k_0 = 1/32$ [153], and for packed capillary columns $k_0 \approx 1/100$ [28].

We turn to the tasks, indicated above, of getting the fastest analysis speed, the narrowest zone widths, and the highest efficiency, which is related to the other parameters via (3.14), (3.15), and (3.17) given k_0, ΔP, η, D_m, D_s, and k'. We shall prove that the possible optimization is limited by the extracolumn spreading (in the injector, the detector, connecting tubing, etc.) which adds to the spreading in the column, thus reducing the apparent efficiency that can be achieved during an experiment. The contribution of the extracolumn spreading θ^2 is defined by the ratio of the volumetric variance of the extracolumn spreading ($\sigma^2_{V,ex}$) to that of the spreading in the column ($\sigma^2_{V,c}$):

$$\theta^2 = \sigma^2_{V,ex}/\sigma^2_{V,c}. \tag{3.20}$$

If it is assumed that the experimental N (measured by the chromatograph) should not be less than 90% of the true column efficiency (N_c), which is equivalent to the assumption of a 5% reduction in K_R (3.14),

$$N = N_c/(1 + \theta^2) \geqq 0.9N_c, \tag{3.21}$$

then $\theta^2 \leqq 0.1$ and $\theta \leqq 0.3$. The different sorts of extracolumn spreading make additive contributions to $\sigma^2_{V,ex}$; thus,

$$\sigma^2_{V,ex} = \sum_j \sigma^2_{V,ex_i}, \quad (3.22)$$

and hence

$$\vartheta^2 = \sum_j \theta^2_i. \quad (3.23)$$

From (3.20) and (3.23) it is clear that the variance contributions of each of the parts of a chromatographic system are related to peak variance due to the column processes by the relation

$$\sigma^2_{V,ex_i} = \theta^2_i \, \sigma^2_{V,c}. \quad (3.24)$$

Since

$$V_R = \sigma_{V,c} N_c^{\frac{1}{2}} = \sigma_{V,ex} N_c^{\frac{1}{2}} = (\tau\varepsilon/4\theta) d_c^2 (1 + k')L,$$

then

$$d_c \geqq \left[\frac{4\sigma_{V,ex} N_c^{\frac{1}{2}}}{\pi\varepsilon L(1 + k')\theta}\right]^{\frac{1}{2}} = \left[\frac{4\sigma_{V,ex}}{\pi\varepsilon N_c^{\frac{1}{2}} h d_p (1 + k')\theta}\right]^{\frac{1}{2}}, \quad (3.25)$$

where ε is the porosity of the column, such that $\varepsilon = \varepsilon_0 + \varepsilon_p$, ε_0 is the interparticle porosity, and ε_p the proportion of pore volume; ε is 0.7-0.88 for packed columns [154], 0.85-0.9 for packed capillary columns with irregular packing [28], and 1 for open tubular columns; $d = d_p$ is the diameter of a sorbent grain for packed columns and $d = d_c$ the column diameter for open tubular capillaries; and $L = Nhd_p$.

It is clear from (3.25) that the lowest allowable column diameter is influenced by two factors: the necessary efficiency N_c for a given k' and L, and the extracolumn spreading.

In order to look at the optimization with respect to the above criteria we use (3.19) to write expressions for t_R (the analysis time), σ_V (the volume standard deviation for the peak), and N, viz.,

$$t_R = \frac{L}{u}(1 + k') = \frac{N^2 h^2 \eta (1 + k')}{k_0 \Delta P} = \frac{N d^2 (1 + k')}{D_m} \frac{h}{\nu}, \quad (3.26)$$

$$\sigma_V = \frac{\tau\varepsilon}{4} N^{\frac{1}{2}} h d \cdot d_c^2 (1 + k') = \frac{\pi\varepsilon}{4} N^{\frac{1}{2}} d_c^2 (1 + k')H, \quad (3.27)$$

$$N = \left[\frac{k_0 \Delta P t_R}{h^2 \eta (1 + k')}\right]^{\frac{1}{2}} = \frac{t_R D_m}{d^2 (1 + k')(h/\nu)} = \frac{k_0 d^2 \Delta P}{\eta D_m (h\nu)}. \quad (3.28)$$

These equations show that the minimum t_R and the maximum N (for a given t_R) will be achieved, if the pressure is limited to a fixed value, when $h = h_{min} \equiv h_0$, which corresponds to $\nu = \nu_{opt} \equiv \nu_0$. The product $h\nu$, with ν being determined from (3.5) or (3.8), can also be found from (3.19), i.e.,

$$h\nu = \frac{k_0 d^2 \Delta P}{N \eta D_m}. \tag{3.29}$$

The product $(h\nu)_0$ (strictly, of the values determined for a column prepared in a standard way) unambiguously defines the important operating parameters for the column, i.e., d the sorbent diameter or open tubular column diameter and ΔP the pressure drop, thus given η, D_m, and N_c:

$$d^2 \Delta P = \left[\frac{(h\nu)_0}{k_0}\right](\eta D_m) N_c. \tag{3.30}$$

The expression in the brackets is a dimensionless hydrodynamic constant for a chromatographic column and depends only on how well it is packed (and the k' of the analyte). We shall call this quantity the column impedance and designate it $E_c = (h\nu)_0/k_0$. The whole right-hand side of (3.30) we shall designate $\tilde{E}_0$. This generalized parameter depends on E_c, the column's efficiency N_c, and the hydrodynamic characteristics of the eluent and solute,

$$d^2 \Delta P = E_c (\eta D_m) N_c = \tilde{E}_0, \tag{3.31}$$

and given $h \neq h_0$ and $\nu \neq \nu_0$, then $d^2 \Delta P \neq \tilde{E}_0$, i.e.,

$$d^2 \Delta P = (h\nu/k_0)(\eta D_m) N_c \equiv \tilde{E}. \tag{3.32}$$

Equation (3.30) yields an experimental way of determining E_c because η, d, and D_m are known, while N_c and ΔP can be found experimentally. Equation (3.31) demonstrates that d and ΔP may vary widely and yet leave the column in the optimal regime for chromatography so long as $d^2 \Delta P = \tilde{E}_0$ remains unchanged.

Obviously, the lower E_c is, the lower is the ΔP (for a fixed d) required to keep the column in the optimal regime. The values of E_c for packed, loosely packed capillary, and open tubular columns are given in Table 3.2, the lowest value being for open tubular columns. Interestingly, the generalized parameter $\tilde{E}_0$ is a weak function of temperature and the quality of the solvent, because a change in the temperature of the mobile phase's composition moves the η and D_m in opposite directions.

Thus, a real chromatographic column is characterized by its column impedance E_c, its efficiency N_c, and d (i.e., d_p for packed columns or d_c for open tubular columns). The required column length is determined by N_c, h, and d thus:

TABLE 3.2. Basic Parameters Defining the Performance of Different Types of Column for Capillary Liquid Chromatography

No.	Column type	h_0	ν_0	k_0	E_c^d
1	Packed columns	2.15[a]	3[a]	10^{-3}	6450
2	Loosely packed capillary columns	2.15[b]	3[b]	1/150	970
3	Open tubular columns	0.72[c]	5.51[c]	1/32	127

a) Data obtained from Eq. (3.7).
b) Also from Eq. (3.7); experimental data [24, 28] much worse.
c) For k' = 2.
d) Column impedance as defined in (3.30).

$$L = N_c h d, \tag{3.33}$$

and in the optimal regime by

$$L_0 = N_c h_0 d_0, \tag{3.34}$$

where the optimum value of $d = d_0$ can be found from (3.31):

$$d_0 = \left[\frac{E_0}{\Delta P}\right]^{\frac{1}{2}} = \left[\frac{E_c(\eta D_m)N_c}{\Delta P}\right]^{\frac{1}{2}}. \tag{3.35}$$

Under these conditions the column is at its shortest. Any change of L from L_0 takes the column out of the optimal regime, since it leads to $h > h_0$ and $\nu \neq \nu_0$.

Thus to get the maximum sensitivity the reduced velocity and the generalized parameter $\tilde{E}$ must be kept at their minimum values, i.e., $h = h_0$ and $\tilde{E} = \tilde{E}_0 = d_0^2 \Delta P$. Here d should be minimized by allowing ΔP to grow (in which case the column will be at its shortest as will the HETP).

The question then arises as to whether, for a given d and a technically possible pressure drop, i.e., $\Delta P > \tilde{E}_0/d^2$, a departure from Eq. (3.31) is better or a change in the pressures to bring the system back to the optimum.

For an optimization with respect to analysis sensitivity the answer is unambiguous. Since in order to get the minimum σ_V it is necessary to have the smallest HETP, then (3.31) must be satisfied, i.e., a decrease in d must be met by an increase in pressure drop. If we also want a small t_R (3.26) or a large $\tilde{N}$ for a given t_R (3.28), then raising ΔP above the necessary $\Delta P = \tilde{E}_0/d^2$ is beneficial only insofar as it reduces $h^2/\Delta P$.

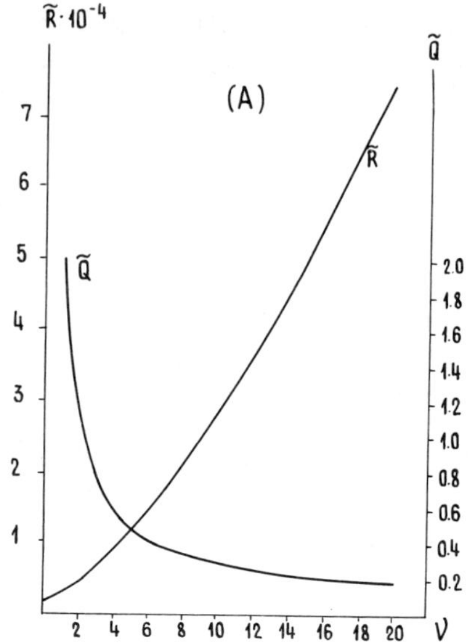

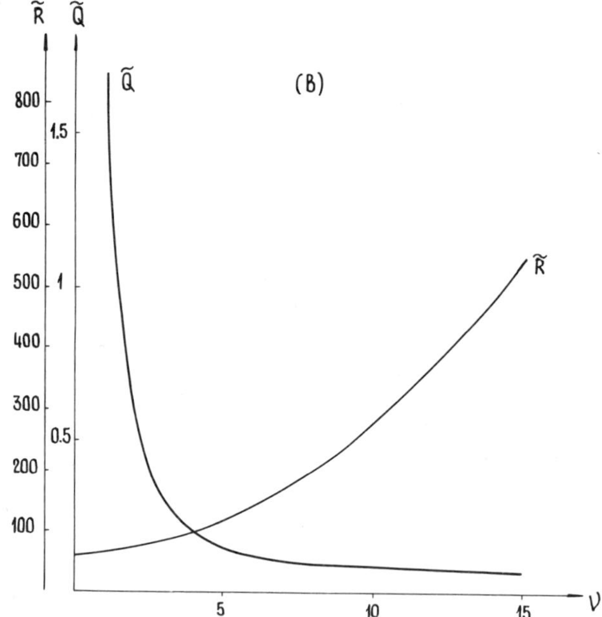

Fig. 3.3. Generalized optimization parameters $\tilde{R} = d^2\Delta P/N_C D_m \eta$ and $\tilde{Q} = t_0 D_m/d^2 N_C$ versus reduced eluent velocity (A) for packed microbore columns, and (B) for open tubular columns. For the packed columns B = 2, A = 1, C = 0.05, and for the open tubular columns B = 2, C = 0.066. η = $5\cdot10^{-4}$ Pa·sec, $D_m = 10^{-9}$ m^2/sec.

We shall now consider the situation in which $h > h_0$ and $\nu \neq \nu_0$. By transforming (3.30) we can, using (3.5), obtain the following expression for packed columns:

$$\tilde{R} = \frac{d^2 \Delta P}{\eta D_m N_c} = \frac{(B + A\nu^{1/3} + C\nu^2)}{k_0} \tag{3.36}$$

and an analogous expression which includes the time taken for the unretained solute ($k' = 0$) to pass through the column, t_0, i.e.,

$$\tilde{Q} = \frac{t_0 D_m}{d_p^2 N_c} = \frac{B}{\nu^2} + \frac{A}{\nu^{2/3}} + C. \tag{3.37}$$

Figure 3.3 contains the curves for the dimensionless parameters $\tilde{R}$ and $\tilde{Q}$ versus the reduced eluent velocity, curves which are important for optimization. These curves establish a relationship between d_p, ΔP, N_c, and t_0 (for given η and D_m). They can be used to obtain any of the above given values for the other three. To do this, take a velocity v' and find the points $\tilde{R} = \tilde{R}(\nu')$ and $\tilde{Q} = \tilde{Q}(\nu')$ on the respective curves. These points will correspond to the same d_p. Given an unconstrained pressure and a fixed d_p, (3.36) and (3.37) can be used to find the minimum analysis time and the maximum efficiency according to (3.26) and (3.28).

Other dimensionless parameters can also be used for optimizations. Bristow and Knox [150] introduced the dimensionless number $E = h_0^2/k_0$, which they called the "separational impedance" and which they use to characterize a column. This parameter is related to t_0, N_c, ΔP, and η, given $h = h_0$, thus

$$E = \frac{h_0^2}{k_0} = \frac{t_0 \Delta P}{N_c^2 \eta}. \tag{3.38}$$

The column impedance E_c we suggest appears to be a better description of a chromatographic system's performance because it is directly related to the generalized optimization parameter $\tilde{E} = d_p^2 \Delta P$. However, the parameter $\tilde{P} = t_0 \Delta P / N_c^2 \eta = (B/\nu + A\nu^{1/3} + C\nu)^2 / k_0$, which is analogous to $\tilde{R}$ and $\tilde{Q}$, may also be used for the optimization together with $\tilde{R}$ and $\tilde{Q}$.

3.3. EXTRACOLUMN SPREADING AND OPTIMIZING THE CHROMATOGRAPHIC SYSTEM

Equation (3.25) indicates that the value of the extracolumn dispersion places restrictions on the diameter of the column.

The main contributions to $\sigma_{V,ex}^2$ are the dispersions in the injector cell, the measuring cell of the detector, and the connecting capillaries. Moreover, the rapidity of the chromatographic analysis is limited by the detector's time constant and the A/D converter's sampling frequency. The analysis mass sensitivity is, in turn, limited by the sample capacity of the column.

We shall now consider each of these factors in more detail.

3.3.1. Dispersion in the Injector Cell

Following Colin et al. [155], we give the volumetric variance due to the dispersion in the injector cell as

$$\sigma^2_{V,s} = V_S^2/K^2, \tag{3.39}$$

where V_S is the sample volume, and K is a constant related to the design of the injector. For a rectangular sample distribution in the injector $K^2 = 12$, usually $K^2 = 4$.

The maximum volume of the test sample is

$$(V_S)_M = \sigma_{V,s}K = \theta_S\sigma_{V,s}K = \theta_S K\frac{\pi\varepsilon}{4}\, d_c^2\, hd(1 + k)N_c^{\frac{1}{2}}. \tag{3.40}$$

Given that K = 2, $\varepsilon = 0.7$, $\theta_S = 0.22$, and $k' = 0$, we have

$$(V_S)_M = 0.28\, d_c^2 hdN_c^{\frac{1}{2}}. \tag{3.41}$$

It is clear from (3.41) that the maximum sample volume is proportional to $N_c^{\frac{1}{2}}$ and to d_c^2 and d_p for packed columns or to d_c for open tubular capillaries. The maximum permissible sample volume is given in Fig. 3.4 as a function of the diameter of a packed column for various efficiencies.

It should be said that the reduction in the sample volume when the column diameter is reduced does not, in itself, lead to a decreased analysis sensitivity if $V_S = (V_S)_M$ when a concentrational detector is used. In fact, the decrease in the sample mass is cancelled out by the decrease in the sample dilution within the column, i.e.,

$$\frac{C_M}{C_0} = \frac{(V_S)_M N^{\frac{1}{2}}}{\sqrt{2\pi}V_R} = \frac{4(V_S)_M}{\pi\sqrt{2\pi}\varepsilon\, d_c^2 hdN_c^{\frac{1}{2}}(1 + k')} = \frac{\theta_S K}{\sqrt{2\pi}}. \tag{3.42}$$

Hence if there is no restriction on the input sample's concentration C_0, then the peak concentration C_M will be independent of the column diameter, i.e.,

$$C_M = C_0\frac{\theta_S K}{\sqrt{2\pi}}. \tag{3.43}$$

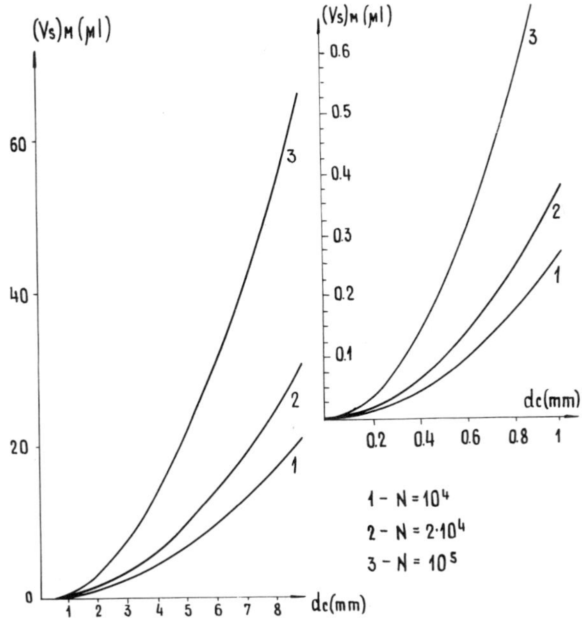

Fig. 3.4. Maximum permissible sample volume as a function of diameter for packed columns with various efficiencies. $(V_S)_M = 0.275 d_c^2 h d_p N^{\frac{1}{2}}$, h = 2, d_p = 5 μm.

3.3.2. Sample Mass

In order to avoid nonlinear sorption isotherms the loading in a packed column should be less than 0.2 mg per gram of sorbent. Hence the concentration in the sample should be no more than 8-10% [31], which means that for packed columns the sample mass q_M is limited thus

$$q_M \leqq 0.08 (V_S)_M \rho_S, \qquad (3.44)$$

where ρ_S is the sample density and can be approximated to 1 g/cm³. For analyte solubilities of less than 30% the concentration must be reduced further and so the mass will be a smaller fraction of $(V_S)_M \rho_S$ than that indicated in (3.44).

The size of the sample mass determines both the preparative capacity of the column and the overall sensitivity given a mass-sensitive detector. For packed columns with h = 2 we have from (3.41)

$$q_M \leqq 0.04 d_c^2 d_p N_c^{\frac{1}{2}} \rho_S. \qquad (3.45)$$

In view of the relatively small volume of the adsorbent layer in open tubular columns their permissible sample mass is severalfold smaller than those for packed columns, i.e.,

$$q_M \leqq 0.015(V_S)_M\rho_S = 0.1d_c^3N_c^{\frac{1}{2}}\rho_S. \qquad (3.46)$$

In the case of a dilute sample and a column with a sorbent that can sorb the sample with $k' \gg 1$ it is possible to increase the volume of the sample, i.e., $V_S >> (V_S)_M$, if the sample is in a solvent for which the k' of the analyte is less than that in the eluent, i.e.,

$$V_S/(V_S)_M = k_s'/k_e'. \qquad (3.47)$$

3.3.3. Dispersion in the Detector Cell

Following Sternberg [156] the volumetric variance due to the dispersion in a detector cell constructed as an ideal mixer is

$$\sigma^2_{V,d} = V_d^2, \qquad (3.48)$$

where V_d is the volume of the cell. Experimental evidence (see Section 5.1) suggests that at large flow rates the detector cell does act as an ideal mixer. Guiochon has suggested [31] that

$$(V_d)_M = (V_S)_M/K, \qquad (3.49)$$

and for K = 2,

$$(V_d)_M = (V_S)_M/2, \qquad (3.50)$$

i.e., the volume of the detector cell must be less than half that of the sample, as given by (3.40).

3.3.4. Dispersion in the Connecting Capillaries

The volumetric variance due to spreading in a capillary with radius r and length ℓ is, following Taylor [152],

$$\sigma^2_{V,con} = \pi r^2\ell F/24D_m, \qquad (3.51)$$

where the volumetric velocity F is related to the linear velocity of the mobile phase in the column by

$$F = \pi\varepsilon d_c^2\, u/4. \qquad (3.52)$$

If we assume that $\theta_{con} = 0.1$, then

$$\sigma^2_{V,con} = \theta^2_{con}\cdot\sigma^2_{V,c} = 0.01\sigma^2_{V,c}, \qquad (3.53)$$

and

$$r^4 \ell < 0.24 D_m \varepsilon d_c^2 h d (1 + k') t_R. \quad (3.54)$$

We can draw an important conclusion from (3.54), viz., that reducing the diameter of the column while retaining the capillary length ℓ does not require a proportional reduction in the capillary diameter, i.e.,

$$r_1/r_2 = (dc_1/dc_2)^{\frac{1}{2}}. \quad (3.55)$$

Hence we can see from this equation that reducing the column diameter by four only requires a twofold reduction in the capillary diameter. This point has great significance when constructing microbore columns, particularly since it opens the way for postcolumn derivatization.

3.3.5. The Detector System's Time Constant

The time constant τ is crucial to the rapidity of the analysis, thus setting certain requirements for the detector, on the one hand, and limiting the rate at which the chromatographic zones can flow through the detector, on the other.

It is believed that τ must be considerably smaller than the time standard deviation of the chromatographic zones:

$$\tau < 0.2\,\sigma_{t,c}. \quad (3.56)$$

Suppose we write the maximum tolerable time constant $(\tau)_M$ as a function of N_c as follows:

$$(\tau)_M \leqq \theta_\tau \frac{t_R}{N_c^{\frac{1}{2}}} = \frac{0.1 t_R}{N_c^{\frac{1}{2}}} \quad (3.57)$$

and substitute $t_R = N^2 h^2 \eta (1 + k')/k_0 \Delta P$; we then get $(\tau)_M$ as a function of the experimental parameters:

$$(\tau)_M = \frac{0.1 N_c^{3/2} h^2 \eta (1 + k')}{k_0 \Delta P}. \quad (3.58)$$

By using (3.57) and (3.26) we can see how $(\tau)_M$ varies for various N_c and d_p for fixed ΔP (Fig. 3.5). It can be seen that reducing N_c and d_P takes $(\tau)_M$ below 0.1, and hence only an oscilloscope or computer can be used for data collection. On the other hand, if we use a detector with a large τ, ignoring (3.58), then we suffer a significant loss of N with respect to N_c, as shown in Fig. 3.6. This figure was generated using

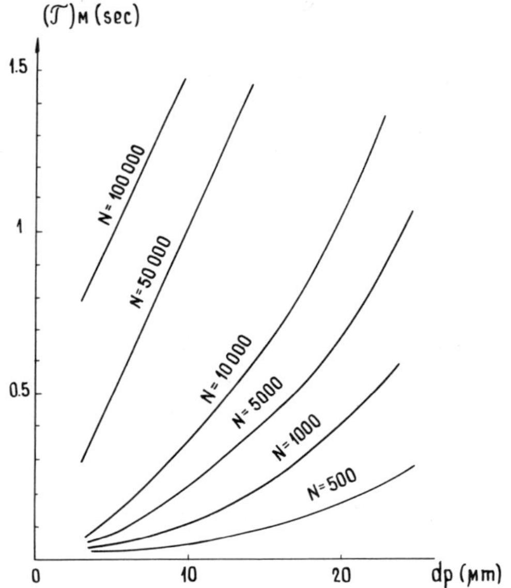

Fig. 3.5. Maximum permissible detector time constant versus column diameter and column efficiency.

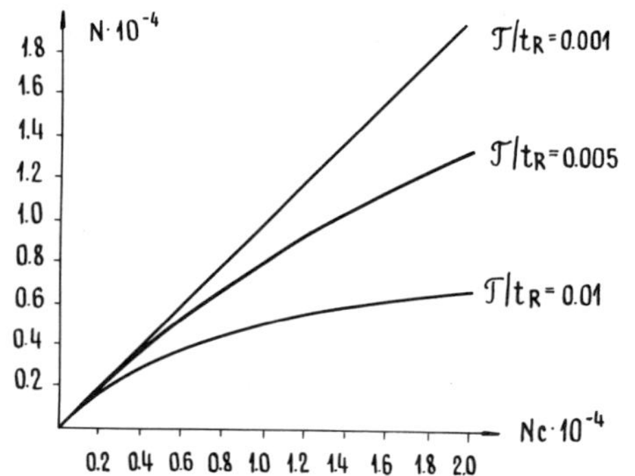

Fig. 3.6. Measurable efficiency N versus real column efficiency N_C for various τ/t_R ratios.

$$N = \frac{1}{1/N_C + [(\tau)_M/t_R]^2}, \tag{3.59}$$

which follows from (3.21) and (3.57).

3.4. OPTIMIZATION WITH RESPECT TO ANALYSIS SPEED

An expression for the analysis time t_0 of an unretained substance ($k' = 0$) is

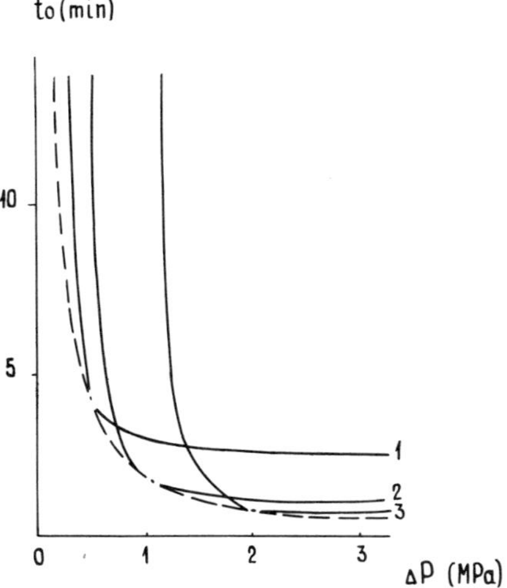

Fig. 3.7. Exit time of unretained peak ($k' = 0$) versus pressure for columns with varying stationary phase sorbent diameters in packed columns using Fig. 3.3. $t_0 = \tilde{Q}(\nu_0)d_p^2N_c/D_m$; $\Delta P = \tilde{R}(\nu_0)\times \eta D_m N_c/d_p^2$. 1) $d_p = 10$ µm; 2) $d_p = 7$ µm; 3) $d_p = 5$ µm.

$$t_0 = \frac{N_c^2h^2\eta}{k_0\Delta P} = \frac{N_c d^2}{D_m}\frac{h}{\nu}. \qquad (3.60)$$

Here ΔP and d are defined in (3.32). Clearly, the shortest t_0 for a given ΔP is achieved when $h = h_0$, for which

$$\tilde{E} = \tilde{E}_0 = (h_0\nu_0/k_0)(D_m\eta)N_c.$$

If ΔP is raised for a fixed d, then t_0 will fall because $h^2/\Delta P$ falls. It can be shown, using (3.60), that for packed columns

$$\frac{h^2}{\Delta P} = \frac{k_0 d^2}{N_c\eta D_m}\left(\frac{B}{\nu^2} + \frac{A}{\nu^{2/3}} + C\right). \qquad (3.61)$$

Since ν rises with ΔP so will $h^2/\Delta P$.

The way t_0 depends on ΔP is illustrated in Fig. 3.7 for packed columns; the dashed line corresponds to $d_{p_0}^2\Delta P = \tilde{E}_0$. Any movement away for the optimal value of d_p leads to a rise in t_0 over $t_0 = t_0(d_{p_0})$. This is used to determine how t_0 depends on d_p (Fig. 3.8). For $d_{p_0}^2\Delta P = E_0$ we get $h/\nu = h_0/\nu_0$.

As can be seen from (3.32) and (3.60), if we fix ΔP and change d_p, then any rise in d_p leads to a fall in h/ν, and vice versa. It

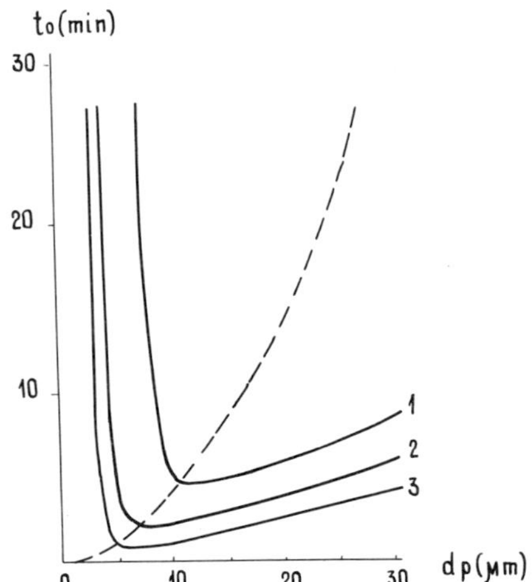

Fig. 3.8. Exit time of unretained peak versus grain diameter of adsorbent for packed columns using Fig. 3.3. $t_0 = \tilde{Q}(\nu_0)d_p^2N_c/D_m$; $\Delta P = \tilde{R}(\nu_0)\eta D_m N_c/d_p^2$. 1) $\Delta P = 0.5$ MPa; 2) $\Delta P = 1$ MPa; 3) $\Delta P = 2$ MPa.

follows from this that $t_0(d_p)$ moves away from a minimum, which is located on the dashed line that corresponds to $d_{p_0}^2\Delta P = \tilde{E}_0$. Any deviation of d_p from d_{p_0} leads to increasing t_0, which is more rapid and hence more dangerous for $d_p < d_{p_0}$.

It follows from (3.26) and (3.60) that the analysis time may be reduced for a fixed ΔP and d_p by reducing η and increasing D_m. This can be achieved by raising the temperature or reducing the viscosity of the solvent, e.g., by using supercritical-fluid chromatography. Characteristically, temperature has a small influence on $\tilde{E}_0$ because, on the one hand, η and D_m are changed in different directions while, on the other hand, a 100°C rise in temperature only changes η and D_m threefold. Going over to supercritical chromatography, meanwhile, changes these parameters five- to tenfold [138].

We now look at some of the other aspects of optimizing with respect to analysis time. The first point to be paid attention to is the danger of reducing d_p below $d_{p_0} = (\tilde{E}_0/\Delta P)^{\frac{1}{2}} = [E_c(\eta D_m)N_c/\Delta P]^{\frac{1}{2}}$. This can occur at relatively low ΔP and large N, i.e., in very efficient columns. The second point is the limitation of the analysis speed by the limitations of the technical characteristics of the detector system, viz., the time constant and the A/D sampling frequency. The specifications of current equipment do not enable us to analyze chromatographically mixtures with more than ten components in less than a minute. On the other hand, completely optimized liquid chromatography requires that the hydraulic and detecting systems work in harmony. Hence in a fast system the whole analysis cycle, including automatic sample injection and data acquisition, has to be completed within 1 min. This sort of apparatus is needed when vast numbers of samples must be processed or single samples processed quickly, as for example in medicine, chemistry, and biotechnology.

So far we have not looked at the speed of a chromatographic analysis from the aspect of the column diameter. Here we distinguish between the following:

1) Heating of the eluent flowing at large velocities in granulated packing. This topic is considered in full in [57]. Eluent heating and the dissipation of the heat through the walls of the column must lead to a viscosity gradient, and hence to a transcolumn gradient in the eluent velocity, which in turn leads to additional spreading of the chromatographic zones. The transcolumn gradient in k', which arises for analogous reasons, leads to the same result. Clearly, a reduction in the diameter of the column will significantly reduce these effects.

2) Obviously, the spreading due to velocity profiles across the column (i.e., the wall effect) and due to nonuniform sorbent packing across the column will be lessened if the diameter is reduced (by a factor proportional to the square of the column diameter).

3) An optimum chromatographic analysis with respect to speed requires that the column be just long enough to get the desired efficiency and hence the column length must be adjusted for each analysis. This is more easily achieved using microbore columns.

4) The flow of eluent corresponding to large ν is much easier to implement in microbore columns than it is in conventional columns.

It should be noted that the requirements of capillary liquid chromatography with respect to the extracolumn dispersion are more easily fulfilled when the analysis is fast but under suboptimal conditions ($h > h_0$) because $(V_s)_M$ and $(V_d)_M$ are increased.

We thus come to the conclusion that small-diameter columns are the most efficient for rapid analysis. The column diameter is bounded from below mainly by the amount of the extracolumn dispersion.

We point out above that in order to computerize a rapid chromatographic analysis an estimate of the speed the A/D converter has to work at to evaluate peaks with small σ_t is required. It is thought that an accurate integration of a Gaussian peak requires about 20 points, though for an asymmetric peak over a noisy baseline 40-50 points are needed. For the latter case a sampling frequency of 400-500 Hz, which is at the limit of commercial devices, is needed, and this enables peaks 100-120 msec wide to be integrated accurately. This would enable us to organize a chromatographic analysis of a mixture with two to five components within 0.5-1 sec. However, a detector with a time constant $(\tau)_M$ of less than 1 msec would then be necessary, but such is at present unavailable. More realistic detectors would have time constants of around 10 msec, yielding analysis times of 3 sec. The requirements on the A/D converter's sampling frequency are thus reduced tenfold, i.e., to 50 Hz. An analysis with $k' = 2$ would thus take 10 sec, while with $k' = 5$ it would take 20 sec. The latter should make it possible to create a microbore chromatograph with an automatic analysis cycle for a 10- to 15-component mixture of 30-60 sec, which includes the automatic injection of the sample and the mathematical processing of the chromatogram. The creation of such a device is now highly desirable.

3.5. OPTIMIZATION WITH RESPECT TO SENSITIVITY (ULTRASENSITIVE ANALYSIS)

The sensitivity of a detector is defined as the ratio of the detector signal δ to the concentration C or flow C·F of the substance in the measuring cell. The relations for a concentration-sensitive and a mass-sensitive detector are

$$\delta_C = k_C C; \ \delta_F = k_F CF, \tag{3.62}$$

where k_C and k_F are the amplification coefficients for both types of detector, F is the flow rate, and C the concentration of the solute through the detector. The concentration in the maximum in the zone, C_M, is related to the sample mass q as follows:

$$C_M = \frac{q}{\sqrt{2\pi}\,\sigma_{V,c}} = \frac{0.7}{d_c^2(1 + k')N_c^{\frac{1}{2}}}\,\frac{1}{H}. \tag{3.63}$$

It can be seen from (3.63) that C_M, as a function of q, is inversely proportional to both the HETP, i.e., H, and the cross-sectional area of the column. This equation indicates that a microbore column with a high performance should be used in order to get an ultrasensitive analysis. The sample mass q is related to the sample volume V_s and the sample concentration C_0:

$$q = C_0 V_s. \tag{3.64}$$

Substituting in the maximum permissible volume of the sample (3.40) then

$$q = C_0 (V_s)_M \cong 0.5 C_0 \sigma_{V,c}. \tag{3.65}$$

It is easy to show that if the injector yields the maximum permissible sample volume, the maximum concentration in a chromatographic zone is always a certain fraction of the sample concentration irrespective of the type of the column, i.e.,

$$C_M = \frac{C_0(V_S)_M}{\sqrt{2\pi}\,\sigma_{V,c}} \cong 0.2C_0. \qquad (3.66)$$

Two sorts of detector are employed in liquid chromatography: a) concentration-sensitive detectors, in which the signal δ_C is proportional to the concentration of the substance in the measuring cell, and b) mass-sensitive detectors, in which the signal δ_F is proportional to the flow of the substance (i.e., the quantity of substance passing through the detector per unit time).

The following relation is characteristic of a concentrational detector:

$$\delta_{CM} = k_C C_M \cong 0.2k_C C_0, \qquad (3.67)$$

which demonstrates that the signal is independent (or weakly dependent) on the column diameter. Meanwhile, the minimum detectable quantity of substance, $\tilde{q}$, is determined by d_c^2, i.e.,

$$q_M = (V_S)_M C_0 = 5C_M(V_S)_M = 50\tilde{C}(V_S)_M \approx 5\tilde{C}d_c^2HN^{\frac{1}{2}}; \qquad (3.68)$$

i.e., it is proportional to the cross-sectional area of the column. Here $\tilde{C}$ is the threshold concentration for the detector, and the factor of 50 in the third expression is associated with the necessity in a quantitative analysis for C_M to be more than tenfold greater than the threshold concentration.

Thus, given a concentration-sensitive detector, reducing the column diameter enables us to reduce the threshold mass of the sample in proportion to the square of the reduction in the column diameter.

For mass-sensitive detectors, the signal is proportional to the square of the column diameter:

$$\delta_{F,M} = k_F C_M F = 0.1k_F d_c^2 u C_0 \quad [\text{for } V_S = (V_S)_M], \qquad (3.69)$$

where u is the linear flow rate of the eluent in the column. Clearly, using a mass-sensitive detector cannot lead to a reduction of threshold detectable mass in a microbore column. The most interesting mass-sensitive detector is the mass spectrometer (see Section 4.3.12). We have shown that the signal of a concentration-sensitive detector does not depend on the diameter or flow rate, nor consequently on the detector cell volume. However, one problem for all concentration-sensitive detectors is the demands made on their detector

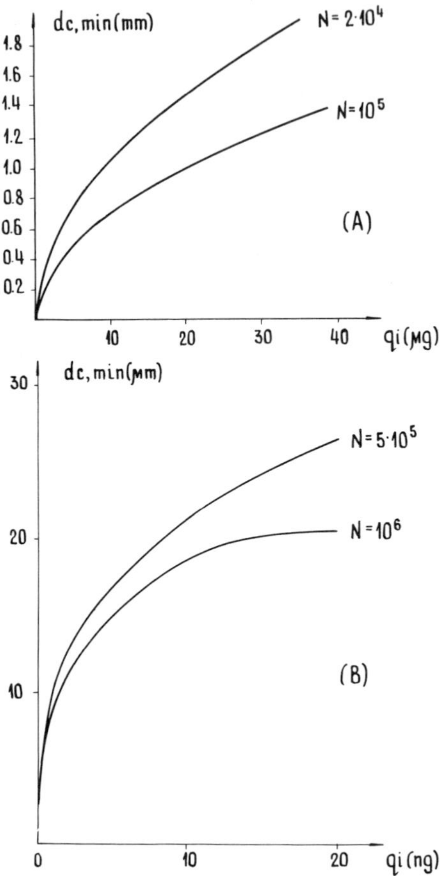

Fig. 3.9. Minimum column diameter versus component mass in sample (β_i = 0.1), given that the maximum permissible sample volume is introduced. A) Packed column, h = 2, d_p = 5 μm, k' = 2; B) open tubular column, h = 1, k' = 2.

cells. For instance, the signal from a photometer detector is related to its optical path length and hence to a certain extent to its cell volume, while a fluorimetric detector's signal is directly determined by its cell volume. In this sense a fluorimetric detector makes the same demands on the separating system as a mass-sensitive detector, i.e., its sensitivity rises with increasing column diameter. Despite the fact that a fluorimetric detector's signal is proportional to d_c^2, its application to microbore chromatography is useful because of the large scope for increasing its concentrational sensitivity k_C. When a mass-sensitive detector is being used, it is natural to ask what the column's smallest diameter is at which, for a constrained mass sensitivity, it can still detect solute.

Assume that a detector has been constructed so that the total mass of the component (q_i) in the sample, which is a fraction β_i of the total sample q ($q_i = \beta_i q$), can enter the detector, and that the maximum permissible sample mass is injected. From (3.44) and (3.46) we have $q_M = 0.08(V_s)_M\rho_s$ for packed columns and $q_M = 0.015(V_s)_M\rho_s$ for open tubular capillary columns. Substituting these into the relevant expressions for $(V_s)_M$ we get (in grams)

$$q_M = 0.08(V_s)_M\rho_s = 0.022N_c^{\frac{1}{2}}d_c^2hd_p(1 + k')\rho_s \tag{3.70}$$

for packed columns, and

$$q_M = 0.015(V_s)_M\rho_s = 0.006N_c^{\frac{1}{2}}d_c^3h(1 + k')\rho_s \tag{3.71}$$

for open tubular capillary columns. Then if the minimum detectable mass of the i-th component (which is a fraction β_i of the total sample mass) is q_i, the column's diameter must be greater than $\tilde{d}_c$ as defined by

$$\tilde{d}_{cpc} = \left[\frac{46\tilde{q}_i}{\rho_s N_c^{\frac{1}{2}}hdp(1 + k')\beta_i}\right]^{1/2}, \tag{3.72}$$

$$\tilde{d}_{cotc} = \left[\frac{167\tilde{q}_i}{\rho_s N_c^{\frac{1}{2}}h(1 + k')\beta_i}\right]^{1/3}. \tag{3.73}$$

In Fig. 3.9, $\tilde{d}_{cpc}$ (A) and $\tilde{d}_{cotc}$ (B) are given versus $\tilde{q}_i$ ($\beta_i = 0.1$) for various N, given the maximum loading in the column. Henceforth we shall assume when investigating a solution with a concentration less than 8% that it was initially concentrated.

These dependences enable us to determine the column diameter $\tilde{d}_c$, assuming the detection of the entire zone of the i-th component. If the mass sensitivity of the detector is specified in units of mass flux, then multiply the expression in the brackets by σ_t (in seconds), which yields the time it takes for a peak fraction with a concentration greater than $0.882C_M$ to pass through the detector.

3.6. SUPEREFFICIENT ANALYSIS

Very high efficiencies ($N > 2 \cdot 10^5$) have been achieved in liquid chromatography that enable mixtures containing hundreds of components to be analyzed (since the peak capacity $n \approx 0.5N^{\frac{1}{2}}$ for $k_n' = 6.4$), or to separate substances with very small $\Delta k'$, such as compounds with and without radioactive isotopic tags. Here capillary technology must be the best because only it can be used to make a single or composite column several meters long without noticeably raising h_0 or E_C above those for columns with conventional lengths

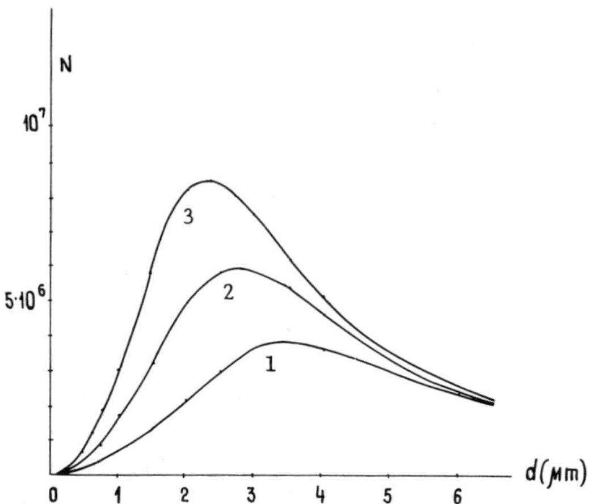

Fig. 3.10. Efficiency of open tubular columns versus diameter for a given time duration. ΔP: 1) 20 MPa; 2) 50 MPa; 3) 100 MPa. $N = t_0 D_m / d_c^2 \tilde{Q}(\nu_0)$; $\Delta P = \tilde{R}(\nu_0)\eta D_m N_c / d_c^2$; $\nu_0 = 0.73$; $\tilde{R}(\nu_0) = 9.1$; $\tilde{Q}(\nu_0) = 2$; $\eta = 5\cdot10^{-4}$ Pa·sec; $D_m = 10^{-9}$ m²/sec; $t_0 = 5$ h.

(10-25 cm). On the other hand, lengthening the column increases its volume, thus making the demands on the volume of the extracolumn system less severe. Nevertheless, other limitations on superefficient chromatographic systems do arise. These are associated with the analysis time t_R, the pressure needed ΔP, and the diameter of the column d_c, which depends on the admissible volume of the extracolumn system.

Let us look at the effects of the first two parameters on N; these can be established from (3.28). It follows from this equation that given t_R, k_0, ΔP, η, and k' an increase in N can be achieved by minimizing $h \rightarrow h_0$. The type of column determines the ratio k_0/h_0^2, which is $2.5\cdot10^{-4}$ for a packed column, $1.7\cdot10^{-3}$ for a packed capillary column, and $6\cdot10^{-2}$ for an open tubular capillary column. These are in the ratio 1:6.8:240, and clearly the inverse ratio 240:35:1 will be obtained for t_R, given a fixed N and ΔP; or for ΔP, given a fixed t_R and N. If, however, we fix ΔP and d, then the ratio for N will be 1:6.5:46 (packed, packed capillary, and open tubular columns, respectively). It is obvious from this that for a superefficient analysis capillary columns, especially open tubular ones, are better than packed columns. We continue and look at the ratios of $d_0 = (\tilde{E}_0/\Delta P)^{\frac{1}{2}}$ for packed capillary and open tubular columns, given a fixed ΔP and N. Given (3.35), the ratio of the d_0 values is proportional to the ratio of the values of $(E_c)^{\frac{1}{2}}$ for the columns and is equal to 6.9. Thus, if a column is packed with a sorbent with $d_p = 10$ μm, then the diameter of an open tubular capillary column must be 1.5 μm, which is not too feasible.

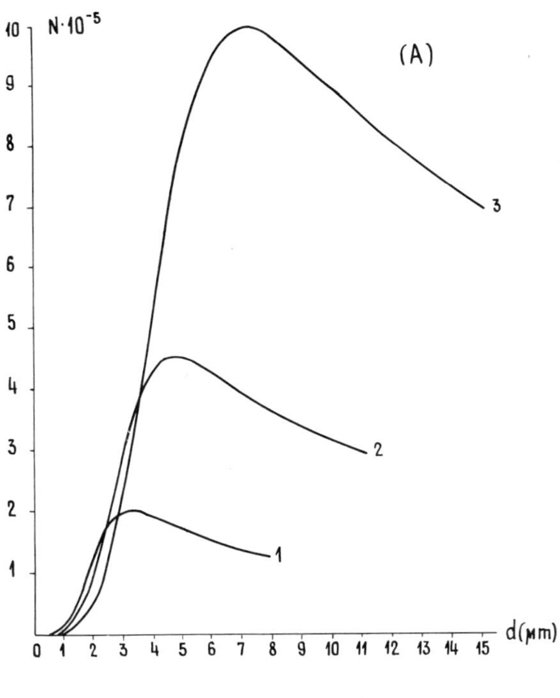

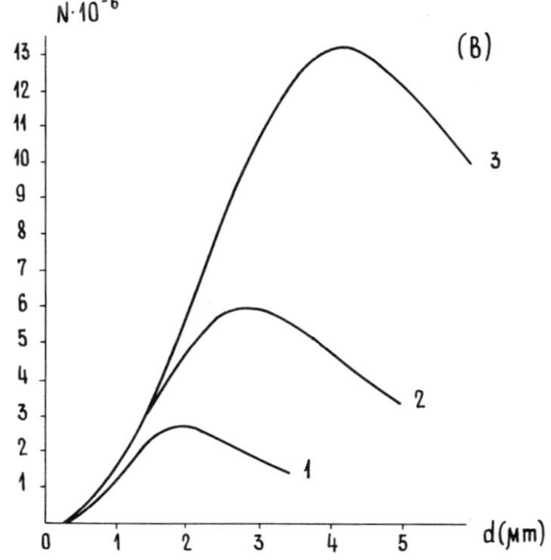

Fig. 3.11. Column efficiency versus grain diameter for A) packed microbore and B) open tubular columns. t_0: 1) 1 h; 2) 5 h; 3) 24 h. For (A), $\tilde{R}(\nu_0) = 6500$, $\tilde{Q}(\nu_0) = 0.7$; the other parameters are as for Fig. 3.10.

The relationship between N and d_c for packed and open tubular capillary columns is shown in Fig. 3.10 for various pressures and a fixed analysis time. In the figure note how N goes through a maximum. Figure 3.11 contains the analogous curve for packed microbore and open tubular columns for a ΔP of 50 MPa and various t_0.

However, the application of open tubular capillaries to superfast analyses is limited by the extracolumn dispersion (the volume of the detector cell). Thus, for a certain N_c it may be better to employ packed columns. To understand this, compare the analysis time for the unretained component, t_0, under the optimal conditions for packed columns, viz., $h = h_0$, $d_p = d_{p0}$, and that for open tubular columns, for which the minimum practically obtainable reduced plate height h is determined by $(V_d)_M^{otc}$ and d_c, i.e.,

$$t_0^{pc} = \frac{(h_0^{pc})^2}{k_0^{pc}} \eta \frac{N_c^2}{\Delta P}, \tag{3.74}$$

$$t_0^{otc} = \frac{(h_0^{otc})^2}{k_0^{otc}} \eta \frac{N_c^2}{\Delta P}. \tag{3.75}$$

According to (3.40) and (3.50) for $k = 2$, $\varepsilon = 1$, $\theta_s = 0.23$, and $k' = 0$, we have

$$h^{otc} = \frac{5.5(V_d)_M^{otc}}{d_c^3 N_c^{\frac{1}{2}}} \tag{3.76}$$

and

$$t_0^{otc} = \frac{30[(V_d)_M^{otc}]^2}{d_c^6 k_0^{otc}} \eta \frac{N_c}{\Delta P}. \tag{3.77}$$

Hence (3.75) and (3.77) indicate that, given a fixed η, N, and ΔP, the ratio t_0^{pc}/t_0^{otc} is

$$\frac{t_0^{pc}}{t_0^{otc}} = \frac{(h_0^{pc})^2(d_c^{otc})^6 N_c}{30\,[(V_d)_M^{otc}]^2} \frac{k_0^{otc}}{k_0^{pc}}. \tag{3.78}$$

From this it is clear that this ratio is independent of ΔP, is linear in N_c, rises (i.e., open tubular columns become more useful) as the detector volume falls, and is very strongly (to the sixth power) dependent on d_c^{otc}. Clearly, at low efficiencies packed columns are better, but at high efficiencies open tubular columns are better. We define an $\bar{N}_c$ below which, for a fixed ΔP, $t_0^{pc} < t_0^{otc}$ and above which $t_0^{pc} > t_0^{otc}$, i.e., the $\bar{N}_c$ at which $t_0^{pc} = t_0^{otc}$. Using (3.78), we have

$$\bar{N}_c = \frac{30[(V_d)_M^{otc}]^2}{(h_0^{pc})^2(d_c^{otc})^6} \frac{k_0^{pc}}{k_0^{otc}}. \tag{3.79}$$

TABLE 3.3. Column Efficiency as a Function of d_c^{otc} and $(V_d)_M^{otc}$ Given $t_0^{pc} = t_0^{otc}$ and $\Delta P^{pc} = \Delta P^{otc}$

$(V_d)_M^{otc}$ (cm^3) \ d_c^{otc} (cm)	10^{-4}	$2 \cdot 10^{-4}$	$3 \cdot 10^{-4}$	$5 \cdot 10^{-4}$	10^{-3}	$2 \cdot 10^{-3}$	$3 \cdot 10^{-3}$
10^{-6}	$2.4 \cdot 10^{11}$	$3.8 \cdot 10^{9}$	$3.3 \cdot 10^{8}$	$1.5 \cdot 10^{7}$	$2.4 \cdot 10^{5}$	$3.8 \cdot 10^{3}$	$3.3 \cdot 10^{2}$
10^{-7}	$2.4 \cdot 10^{9}$	$3.8 \cdot 10^{7}$	$3.3 \cdot 10^{6}$	$1.5 \cdot 10^{5}$	$2.4 \cdot 10^{3}$		

The results can be seen in Table 3.3.

As can be seen from Table 3.3, a 10-μm open tubular column begins to have a shorter minimum time once N_C exceeds $2.4 \cdot 10^5$, given a detector volume of 1 nl, and $2.4 \cdot 10^3$ (i.e., practically the whole efficiency range), given a detector volume of 0.1 nl. Similar considerations can be used to choose which column type is best for a concrete superfast analysis.

It follows from (3.29) that reducing the viscosity favors a significant (proportional to $\eta^{\frac{1}{2}}$) increase in N_C for a constant ΔP and t_R, or a decrease (proportional to η) in the analysis time for a constant ΔP and N. Hence, raising the temperature of the column or using supercritical-fluid chromatography instead of conventional liquid chromatography is a good way of raising the efficiency.

It is clear that the rationality of a superefficient system is determined by the duration of the analysis time, and thus on the productivity of the system N/t_R, i.e.,

$$\frac{N}{t_R} = \frac{k_0}{h^2}\,\frac{\Delta P}{N}\,\frac{1}{\eta(1 + k')}. \tag{3.80}$$

This indicates that the highest productivity for a fixed ΔP comes when $h = h_0$ and rises with a fall in viscosity. In order of increasing productivity we have packed, packed microbore, and open tubular columns.

3.7. CONCLUSION

We have considered three sorts of optimization, i.e., with respect to speed of analysis, to sensitivity, and to efficiency. Optimization for analysis speed at a given pressure is best done using the separational impedance of Bristow and Knox [150] ($E = h_0^2/k_0$), i.e.,

$$t_0 = \frac{N_C^2 h^2 \eta}{h_0 \Delta P} = E\frac{N^2 \eta}{\Delta P}, \tag{3.81}$$

while column optimization for ΔP and d is easiest with the column impedance as defined in (3.31).

Speed, efficiency, and analysis productivity optimizations when the pressure is not constrained are best done with the dimensionless number $\tilde{Q}$ (3.37), i.e.,

$$t_0 = \tilde{Q}\frac{d^2 N_C}{D_m} = \frac{h}{\nu}\,\frac{d^2 N_C}{D_m}. \tag{3.82}$$

Only microbore columns, and at the limit of their performance open tubular columns, can exploit the potentials of liquid chromatography, i.e., very fast, ultrasensitive (for a threshold sample mass), and superefficient (with regard to N) analyses. These very good results are due directly to the small column diameters (for the ultrasensitive analyses with concentration-sensitive detectors) or to technological peculiarities of capillary columns such as the good heat dissipation during rapid analyses and the ability to make long composite or single columns without noticeably increasing the HETP for a superefficient analysis. On the other hand, open tubular capillary columns are limited in application because of the impossibility of reducing the volume of the injector or detector cell below certain limits. This need for precise microinjectors and micropumps that can handle small eluent flows, and the complexity of getting detectors with the requisite concentrational sensitivity, given measuring cells with tiny volumes, are the technical difficulties that face the manufacturers of microbore liquid chromatographs.

Besides the unmatched optimal regimes, microbore technology also offers significant savings due to the reduced expenditure on costly sorbents and solvents, and hence enables exotic or toxic eluents and expensive specialized sorbents to be utilized.

Bringing down the volumes of eluent makes it possible to use capillary columns with detectors for which large quantities of eluent are impractical, such as mass spectrometers, or to use the whole of the test sample as the eluent and hence realize vacant chromatography.

Finally, it is worth mentioning that mainly due to its miniaturization capillary liquid chromatography is very convenient for applying innovations such as fused-silica capillary columns, laser fluorimetry, and the use of supercritical fluids as eluents.

Chapter 4

EQUIPMENT FOR CAPILLARY LIQUID CHROMATOGRAPHY

4.1. SPECIFICATIONS

The advantages of capillary liquid chromatography with respect to efficiency, speed, economy, and sensitivity of analysis can only be fully realized if the ancillary equipment is miniaturized too. Primarily, the volumetric flow rate F, which is related to the linear velocity u of the mobile phase by the following expression, must be reduced:

$$F = \frac{\pi \varepsilon d_c^2 u}{4} = \frac{\pi \varepsilon d_c^2 D_m \nu}{4 d_p}, \tag{4.1}$$

where ε is the overall porosity of the column (the fraction of the volume of the column occupied by the pores and the interparticle channels), $\nu = ud_p/D_m$ is the reduced eluent velocity, and D_m is the diffusion coefficient. The same range of linear eluent velocities may be used in capillary chromatography as is used in conventional chromatography, i.e., u = 0.01-5 cm/sec. However, the corresponding pump deliveries of eluent are reduced by d_c^2, as follows from (4.1). Similarly, the quantity of solvent needed to carry out one analysis is also reduced, i.e.,

$$V_T = \frac{\pi \varepsilon d_c^2 L(1 + k')}{4}. \tag{4.2}$$

The maximum quantity of solvent expended in one analysis may be estimated from (4.2) as the retention volume of the compound with a great affinity to the stationary phase, e.g., k' = 10. Reducing the diameter of the column stiffens the limits on the allowable volumes of the injector, connecting tubes, and detector cell, and for a rapid analysis the response time of the registration system, i.e., the detector itself, the amplifier, and the recorder. We shall consider a chromatograph to be good if its efficiency is only reduced by 15% due to extracolumn peak broadening, viz.,

$$\theta^2 = N_c/N - 1 \leqq 0.15, \tag{4.3}$$

where N is the apparent efficiency and N_c is the actual column efficiency. There are three main sources of the reduction in the efficiency: band spreading in the detector cell, which can be characterized by a volumetric variance $\sigma^2_{V,d}$; spreading in the injector, with a variance of $\sigma^2_{V,s}$; and distortion of the peaks due to the finite time constant of the detector, which can be described by a variance $\sigma^2_{t,\tau} \cdot F^2$. The spreading in the connecting tubes may be ignored since Scott and Kucera [14, 52] have described ways of connecting a microbore column to the detector cell and injector without intermediate tubing. The allowable volumes of the sample $(V_s)_M$ and detector cell $(V_d)_M$ and the maximum time constant $(\tau)_M$ acceptable for a given column length can be estimated from the following equation, which is derived from the definition of N and the additivity law for variances, i.e.,

$$\theta^2 \approx \frac{\sigma^2_{V,d} + \sigma^2_{V,s} + \sigma^2_{t,\tau}F}{\sigma^2_{V,c}} = \theta^2_d + \theta^2_s + \theta^2_\tau, \tag{4.4}$$

where

$$\sigma_{V,c} = \sigma_{t,c}F = \pi\varepsilon(d_c^2/4)(1 + k')\sqrt{Ld_ph}$$

is the standard deviation of the peak (in units of volume) and describes the spreading in the column, h is the reduced height equivalent to a theoretical plate, L is the length of the column, and d_p is the diameter of a sorbent particle.

We shall consider that the sources of extracolumn spreading contribute equally to θ^2, and since $\theta^2 \leqq 0.15$ for a good chromatograph, each contribution must be less than 5%, i.e.,

$$\theta^2_s = \theta^2_d = \theta^2_\tau \leqq 0.05.$$

It is reasonable to assume that at large flow rates the injector and detector cell are ideal mixers (see Section 5.1), i.e., $\sigma^2_{V,d} = V^2_d$ and $\sigma^2_{V,s} = V^2_s$. The other limiting case is plug flow, for which the variances may [155, 156] be calculated from the second statistical moment of a rectangular distribution, i.e., $\sigma^2_{V,d} = V^2_d/12$ and $\sigma^2_{V,s} = V^2_s/12$.

These relations together with (4.4) enable us to obtain the following expression for the maximum permissible volumes of the injector and measuring cell:

$$(V_s)_M = (V_d)_M = \theta K \frac{\pi \varepsilon d_c^2 (1 + k')\sqrt{Lhd_p}}{4}, \tag{4.5}$$

where $K = 1$ for large u, and $K = \sqrt{12} = 3.46$ for plug flow, which apparently is the case for quite low linear liquid velocities at the minimum of the dependence HETP versus velocity. The distortion of the peaks due to the inertia of the detector leads to a loss of efficiency θ_τ. This is characterized by the limiting time constant $(\tau)_M$, which is related to the actual efficiency of the column, as follows:

$$(\tau)_M = \theta_\tau \sigma_{t,c} = \theta_\tau \frac{h}{\nu} \frac{N_c^{\frac{1}{2}} d_p^2 (1 + k')}{D_m} = \theta_\tau \frac{t_R}{\sqrt{N_c}}, \tag{4.6}$$

where $\sigma_{t,c}$ is the standard deviation (in units of time) for the spreading in the column, and t_R is the analysis time.

The parameters of liquid chromatography for packed microbore columns with different diameters are tabulated in Table 4.1. They were calculated using Eqs. (4.1)-(4.6) for an analysis with the maximum efficiency, i.e., at the minimum of the HETP versus velocity curve, viz., $h_0 = 2$, $\nu_0 = 3$. The efficiency of the column can be calculated from

$$N = \frac{L}{hd_p}. \tag{4.7}$$

The parameters for packed columns and rapid analyses (Table 4.2) have also been calculated using (4.1)-(4.7). However, in this case the reduced velocity (of the mobile phase) was calculated from the maximum pressure in the chromatograph $\Delta P = 40$ MPa using

$$\nu = \frac{k_0 d_p^3 \Delta P}{L \eta D_m}, \tag{4.8}$$

where k_0 is a dimensionless number describing the permeability of the column ($k_0 \cong 10^{-3}$ for packed columns with regular packing), and η is the viscosity of the mobile phase.

The value of the reduced HETP can then be calculated from Knox's equation

$$h = \frac{2}{\nu} + \nu^{1/3} + 0.1. \tag{4.9}$$

Since open tubular capillary columns only become advantageous for $N_c > 2 \cdot 10^5$, as shown in Chapter 3, we used the following initial data for calculating their parameters (Table 4.3): $N_c = 10^6$, $h_0 = 0.8$, $\nu_0 = 5$, which corresponds [29] to the minimum of $h(\nu)$. The parameters were generated using (4.1), (4.2), and (4.5)-(4.7) (d_p being replaced by d_c) and the characteristic parameter $\varepsilon = 1$.

TABLE 4.1. Calculated Parameters for a Superefficient Chromatograph Using Packed Columns with Different Diameters (h_0 = 2, ν_0 = 3, d_p = 5 μm)

d_c (mm)	L (cm)	N_c^{-3} (a)	$(V_s)_M$ (b) (μl)	$(V_d)_M$ (b) (μl)	F (c) (μl/min)	V_T (d) (ml)	$(\tau)_M$ (e) (sec)
4			7.3	7.3	340	10.4	
1	10	10	0.45	0.45	21.3	0.65	0.37
0.5			0.11	0.11	5.3	0.16	
0.2			0.02	0.02	0.85	0.03	
4			11.5	11.5	340	26	
1	25	25	0.73	0.73	21.3	1.6	0.58
0.5			0.18	0.18	5.3	0.4	
0.2			0.03	0.03	0.85	0.08	
4			16.3	16.3	340	51.8	
1	50	50	1.0	1.0	21.3	3.25	0.82
0.5			0.25	0.25	5.3	0.8	
0.2			0.04	0.04	0.85	0.13	
4			22.9	22.9	340	104	
1	100	100	1.43	1.43	21.3	6.5	1.2
0.5			0.35	0.35	5.3	1.63	
0.2			0.06	0.06	0.85	0.25	

Note. a) Equation (4.7). b) Equation (4.5); k' = 0, ε = 0.75, θ = 0.22: K = 3.46 (plug flow). c) Equation (4.1); $D_m = 10^{-5}$ $cm^2 \times sec^{-1}$. d) Equation (4.2); k' = 10. e) Equation (4.6).

The parameters in Tables 4.1-4.3 demonstrate how serious the problems are that accompany the creation of capillary liquid chromatographs. Even in the simplest case (d_c = 1 mm) a fast analysis necessitates a rapidly acting detector [$(\tau)_M \leqq 0.05$ sec] with a measuring cell of 0.3-0.5 μl, and the ability reproducibly to inject a sample of the same volume into the column and provide a stable eluent delivery of 20-300 μl/min in an isocratic or gradient regime. Equipment compatible with packed columns 1 mm in diameter have now almost been completely developed and are now on the market, even though, as we shall see, some of the problems still need development. We are still unable to maintain the sensitivity of concentrational detectors as the volume of the detector cell is brought down or to reproducibly form gradients for final solvent concentrations in the ranges 0-10% and 90-100%.

The situation is worse for the ancillary equipment needed for packed columns 200-300 μm in diameter (fused-silica columns) and worse still for open tubular columns. As Knox and Gilbert [22, 23]

TABLE 4.2. Calculated Parameters for a Fast Chromatograph Using Packed Columns with Different Diameters (ΔP = 40 MPa, d_p = 5 μm)

d_c (mm)	L (cm)	$\nu^{(a)}$	$h^{(b)}$	$N_c \cdot 10^{-3(c)}$	$(V_s)_M{}^{(d)}$ (μl)	$(V_d)_M{}^{(d)}$ (μl)	$F^{(e)}$ (μl/min)	$V_T{}^{(f)}$ (ml)	$(\tau)_M{}^{(g)}$ (sec)
4					4.3	4.3	5540	10.4	
1	10	49	8.6	2.325	0.27	0.27	346	0.65	0.047
0.5					0.07	0.07	86.5	0.16	
0.2					0.01	0.01	13.8	0.03	
4					5.06	5.06	2215	26	
1	25	19.6	4.76	10.5	0.32	0.32	138.5	1.6	0.137
0.5					0.08	0.08	34.6	0.4	
2					0.013	0.013	5.54	0.08	
4					5.98	5.98	1107	51.8	
1	50	9.8	3.32	30.12	0.37	0.37	69.3	3.25	0.32
0.5					0.093	0.093	17.3	0.8	
2					0.015	0.015	2.77	0.13	
4					7.5	7.5	554	104	
1	400	4.9	2.6	76.9	0.47	0.47	34.6	6.5	0.81
0.5					0.12	0.12	3.65	1.63	
2					0.019	0.019	1.38	0.25	

<u>Note</u>. a) Equation (4.8); $D_m = 10^{-5}$ cm²/sec; $k_0 = 10^{-3}$; $\eta = 10^{-2}$ dyne·sec/cm². b) Equation (4.9). c) Equation (4.7). d) Equation (4.5); θ = 0.22; K = 1 (ideal mixer); k' = 0. e) Equation (4.1). f) Equation (4.2); k' = 10. g) Equation (4.6).

TABLE 4.3. Calculated Parameters for a Superefficient Chromatograph Using Open Tubular Columns ($h_0 = 0.8$, $N_c = 10^6$)

d_c (μm)	$L^{(a)}$ (m)	$F^{(b)}$ (ml/min)	$(V_s)_M^{(c)}$ (nl)	$(V_d)_M^{(c)}$ (nl)	$V_T^{(d)}$ (μl)	$(\tau)_M^{(e)}$ (sec)
60	48	14.4	103	103	249	127
30	24	7.2	12.9	12.9	62.2	31.7
10	8	2.4	0.48	0.48	6.9	3.5

Note. For this calculation d_p was replaced by d_c in Eqs. (4.1)-(4.7). a) Equation (4.7). b) Equation (4.1); $\varepsilon = 1$; $D_m = 10^{-5}$ $cm^2 \times sec^{-1}$. c) Equation (4.5); $\theta = 0.22$; $K = 3.46$; $\varepsilon = 1$; $k' = 0$. d) Equation (4.2); $k' = 10$. e) Equation (4.6).

showed for open tubular columns, they must have diameters less than 10 μm in order to compete successfully with packed columns containing particles 5-10 μm in diameter. The characteristic parameters for this sort of column (Table 4.3) are on the edges of the abilities of current technology, viz., F = 2.5 nl/min and $(V_s)_M = (V_d)_M \leqq 0.5$ nl. Nevertheless, a number of research groups have been able to demonstrate very sensitive detection in volumes less than a nanoliter by applying fluorescence [100, 101], electrochemical [102], photometric [98, 99], or potentiometric [103, 104] detectors. Some success has been achieved in developing methods for injecting subnanoliter volumes [108]. In the following sections we shall be looking at the presently available commercial and laboratory equipment for packed microbore columns with diameters of 0.5-1.0 mm, reviewing the apparatus that has recently been developed for open tubular or loosely packed capillary columns with diameters of 10-60 μm, and considering prospective improvement.

4.2. MICROBORE LIQUID CHROMATOGRAPHS

At present, equipment for packed microbore columns (diameters of 0.5-1.0 mm) is available from a number of instrument makers, and the number of firms that have started to produce equipment compatible with microbore columns is growing rapidly. It is thus impossible to claim that what follows is an exhaustive description of the equipment commercially available. The instrument makers are divided fairly clearly in two. Some firms, such as JASCO (Japan), ISCO (USA), Brownlee Labs (USA), and the Science and Technology Organization of the Academy of Sciences (USSR), produce special-purpose equipment for microbore columns, while other manufacturers, such as LKB Produktor AB (Sweden), EM Science (USA), Gilson (France), Waters (USA), and Kratos (USA), try to unify their chromatographic equipment so that it can serve the microbore (d_c = 1 mm), analytical (d_c = 4 mm), and semipreparative (d_c = 8 mm) technologies at once.

The first tendency leads to small-volume syringe pumps, which can provide stable eluent deliveries with flow rates of 1-1000 μl/min, and specially designed spectrophotometric detectors, which combine fast response, sensitivity, and low band spreading. The second tendency, that of unifying the equipment, is realized by the use of replaceable modules for the detector cell and replaceable liquid heads for reciprocating pumps, which can be used to vary the range of deliveries from 0.5 to 10^5 μl/min. It seems to us that the first of these approaches is the more promising because any universal system is going to be significantly worse in satisfying specifications than a narrowly specialized system.

The main parameters of several microbore chromatographs are given in Table 4.4. Although syringe pumps give a more stable delivery, they cannot provide the top pressures reciprocating pumps can. Meanwhile, the volume of the detector cell of a photometric detector can be shrunk to 0.1-1.0 μl, but only an ingenious cell design will yield a reasonable optical path length of 5-10 mm. The KhZh-1309 microbore chromatograph, which is produced in the USSR, is worth mentioning here. It is, so far, the only system for capillary liquid chromatography with a universal refractive index detector. However, the inability to use pressures above 12 MPa that is attendant on the use of PTFE columns must be considered a flaw of the system. The fastest microbore system is the Familic 300-S because it has the capacity for gradients for pressures up to ΔP = 50 MPa and its recording system has a fast response with a time constant of $(\tau)_M$ = 50 msec.

Many firms, as we have said, while not producing specialized equipment for microbore chromatography, do manufacture interchangeable modules for their detectors and pumps to make them compatible to microbore columns. For instance, a good microbore chromatograph can be built up using the following modules made by Waters: a model 510 or 590 solvent delivery system (F = 1-45,000 μl/min), a model 680 or 720 automatic gradient controller, and a model 440 or 441 UV detector with a replaceable cell (V_d = 1.9 μl and ℓ = 10 cm). This system [158] can, for eluent gradient regimes, provide reproducible peak areas with relative standard deviations of 1.5-2.6%. The very small time constant [$(\tau)_M$ = 25 msec] of the detector makes it amenable for superfast analyses. It is also significant that the length of the optical path, and hence the concentrational sensitivity of the detection, remains the same even for the microbore measuring cell (V_d = 1.9 μl). An isocratic microbore system can

TABLE 4.4. Comparison of Microbore Liquid Chromatographs

Chromatograph name	Manufacturer[a]	Solvent delivery system				
		pump[b]	range of F, μl/min	ΔP (MPa)	stability, %	gradient[c]
Familic 100-N	1	MSP (0.5 ml)	1-30	10	1	none
Familic 300	1	RP (0.5 ml)	10-990	50	1	none
Familic 300-S	1	RP (3)	10-990	50	1	LRMG(1)
LC-5A	2	RP (1)	1-9900	50	1	none
LKB-Microbore	3	RP (2)	10-5000	35	0.3	none
MACS	4	RP (1)	1-4500	40	0.5	none
ISCO-314-V_4 System	5	HPSP (370 ml)	0.15-3300	20	0.1	none
KhZh-1309	6	MSP (1.25 ml)	0.5-30	12	0.5	none
KhZh-1310	6	MSP (1.25 ml)	0.5-30	12	0.5	none
KZhKh-1	6	MSP (2 ml)	1-70	12	0.5	HPMG(2)
Varian 5500 (Microbore operation)	7	RP (1)	10-10,000	42	1	LPMG(I)
μ-LC system A	5	HPSP (50 ml)	0.02-600	70	1	none

Note. a) The names of the firms are given in Table 4.5. b) MSP – cating pump. The number in parentheses is the chamber capacity for pumps. c) LPMG(1) is the low-pressure mixing gradient using one d) UVS – visible UV spectrometer; UVP – fixed-wavelength UV photom- tector. e) MLI – microloop injector; SFI – stationary flow injec- glass-lined stainless steel. –) No information.

Detector				Injecting system		Column	
type(d)	Vd, μl	τ, sec	cell length, mm	injector(e)	Vs, μl	column material(f)	dc, mm
UVS SF	0.3 0.6	1.0	0.5	SFI MLI	0.3	PTFE	0.5
UVS	1.0	1.0	5	MLI	1.0	SS	1.5
UVS	1.0	1.0	5	MLI	1.0	SS	1.5
UVS	0.5	0.2	3	MLI	0.5; 1.0	SS	1.0
UVS	0.8	–	3	MLI	0.5; 1.0	GLSS	1.0
UVS	0.5	0.1		MLI	0.5	SS	1.0
UVS	1.0 0.12	0.5	5 1	MLI	0.2	SS	1.0
RID	0.1	1.0	—	MLI	0.25; 0.5	PTFE	0.5
UVP	0.5 0.15	0.5	2.5 0.5	MLI	0.25; 0.5	PTFE	0.5
UVS	1.5 0.5	1.0	5	SFI	1.0	SS	0.5 1.0
UVS	0.5	0.05	–	MLI	0.5	SS	1.0
UVS	0.5 0.25 0.03	0.05	10 5 1	MLI	0.1	SS	1.0

microsyringe pump; HPSP – high-pressure syringe pump; RP – recipro-
the syringe pumps or the number of pistons for the reciprocating
pump; HPMG(2) is the high-pressure mixing gradient using two pumps.
eter; SF – spectrofluorometric detector; RID – refractive index de-
tor. f) PTFE – polytetrafluoroethylene; SS – stainless steel; GLSS –

TABLE 4.5. Some Equipment Suppliers for Microbore Liquid Chromatographs

No.	Firm (country)	Product
1	JASCO (Japan)	Microbore liquid chromatograph
2	Shimadzu (Japan)	Microbore liquid chromatograph
3	LKB-Produkter AB (Sweden)	Microbore modification of liquid chromatograph
4	EM Science (USA)	Microbore chromatograph
5	ISCO (USA)	Microbore chromatograph, syringe pump, ultrasensitive spectrophotometric detector
6	Science and Technology Organization of the Academy of Sciences (USSR)	Microbore liquid chromatograph; photometer, refractive index, and fluorometer detectors; syringe pumps; gradient devices
7	Varian (USA)	Microbore variant of chromatograph
8	Brownlee Labs (Millipore) (USA)	Gradient system based on syringe pumps
9	Waters (France)	Solvent delivery system based on reciprocating pumps
10	Gilson (France)	Microbore chromatograph with replaceable modules; spectrophotometer and fluorometer detectors
11	Knauer (West Germany)	Modules for microbore chromatographs
12	Perkin–Elmer (USA)	Modules for microbore chromatographs
13	Micrometrics (USA)	Modules for microbore chromatographs
14	Valco (USA)	Rotor dispensers
15	Rheodyne Corp. (USA)	Rotor dispensers, filters
16	Farrand Optical Co. (USA)	Fluorometers
17	Ikatos Anal. Instr. (USA)	Fluorometers
18	Kratos (USA)	Spectrophotometer and fluorometer detectors

also be put together using modules made by Gilson: a model 302 pump with a No. 5 head (F = 5-5000 μl/min), a Rheodyne 7410 injector with loop of 0.5 μl, and one of the following detectors: a HM/Holochrome variable-wavelength spectrophotometer, a IIIB UV spectrophotometer, or a model 121 fluorescence detector. Gilson's absorbance detectors are available with a microbore cell module of 1.3 μl (ℓ = 5 mm), while the fluorometer has a cell volume of 0.6 μl.

Knauer (West Germany) offers similar equipment based on their No. 64.00 reciprocating pump, which has a replaceable head for microbore technology (F = 2-2000 μl/min), and No. 97.00 UV photometer or No. 87.00 spectrophotometer (τ = 150 msec), with replaceable flow cells ($V_d \cong 1$ μl and ℓ = 1 mm).

It is also possible to put together a system using modules from different companies. A very successful system consists of a gradient pump system (the MPLC micropump from Brownlee Labs), a detector from Kratos, either the SF-773 spectrophotometer (V_d = 0.5 μl, ℓ = 3 mm) or the FS-970 fluorometer (V_d = 0.25 μl), and the Rheodyne 7520 injector.

4.3. DETECTORS FOR CAPILLARY LIQUID CHROMATOGRAPHY

As for all other chromatography methods, the capillary technique requires a sensitive detector for observing when the effluent's concentration changes. Besides the requirements listed in Section 4.1 (cell volume and time constant) there are no others that distinguish capillary technology from conventional high-performance liquid chromatography. Hence we shall not go into the classification of detectors, their characteristics, etc., since the general questions have all been considered elsewhere in monographs [159, 160] or reviews [161-163].

4.3.1. Photometric (Absorbance) Detectors

The commercial flow photometers for conventional chromatography have cell volumes of $V_d \geqq 8$ μl, which makes them incompatible with capillary columns. The first absorbance detector suitable for capillary columns was described in 1973 [10, 36]. It worked using a symmetrical double-beam optical system. The flow cell was 0.8 mm in diameter and the optical path 0.16 cm long. The reference and measuring cells were as close together as possible, and hence the same silica window could be used for both, thus minimizing the spectral differences between the cells. The placement of the two chambers also meant that they could both be illuminated by a single condenser (with a mirror objective). The light beam was focused on each of the cells by a convex mirror, which oscillated between two fixed positions with a frequency of 25 Hz. This beam splitter was

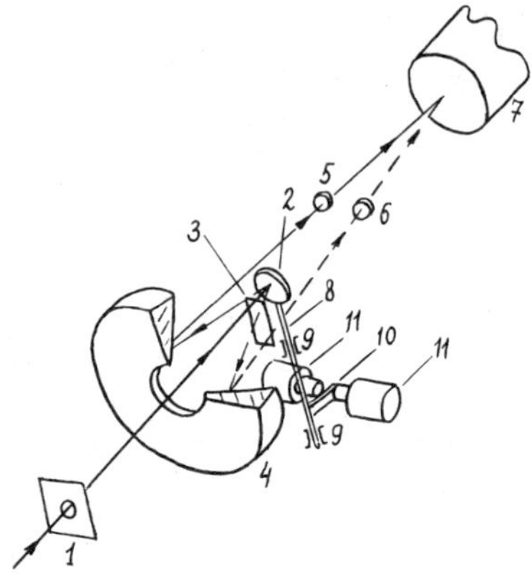

Fig. 4.1. Basic diagram of the photometric part of the MSFP-1 microspectrophotometer. 1) Field diaphragm; 2) vibrating convex mirror; 3) screen; 4) concave mirror; 5) reference cell; 6) measuring cell; 7) photomultiplier; 8) axis; 9) bearing; 10) armature; 11) electromagnet. (Reproduced from [10] with the permission of the authors.)

placed at the focus of a concave mirror, which collimated the monochromatic light. The dimensions of the illuminated section of the cell windows were 0.4 mm × 0.05 mm. The basic setup for the spectrophotometer is shown in Fig. 4.1. The effluent intake to the measuring cell and the eluent intake to the reference cell were brought, via flared polyethylene capillaries (internal diameters 0.3 mm), through grooves in the fluoroplastic bodies of the cells. The dynamic volume of the whole system, including the connecting capillaries, was 2 μl, i.e., a peak width of $4\sigma_V$, given an injection of 0.1 μl of the sample directly into the cell. This design was used in the USSR for the variable-wavelength absorbance detectors in the KhZh-1305 and KhZh-3301 microbore chromatographs, and recently for the SFD-1 detector, which is included in the KZhKh-1 system (see Table 4.4).

Very many instrument makers now produce UV photometers and UV spectrophotometers with optional flowcells, mainly 0.3-1.5 μl in size, for registering effluents from microbore columns. Most use a cylindrical z configuration. The industry leader for optical detectors (photometers and fluorometers) is Kratos. Their Spectroflow 773, which has a microcell with V_d = 0.5 μl, ℓ = 3 mm, and τ = 0.05 sec, is one of the best examples of what can be achieved in photometric detection for microbore columns. Its minimum detectable concentration is 10^{-7} g/ml [164], which is equivalent to 50 pg of sample in the cell volume.

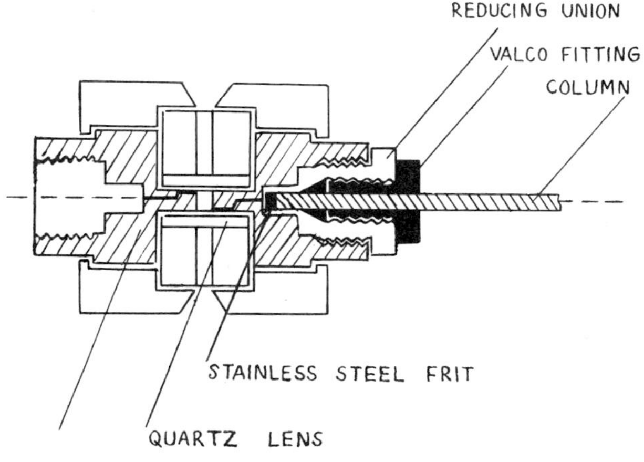

Fig. 4.2. Diagram of flow cell of the Kratos SF-770 spectrophotometric detector and its connections to a steel microbore column. (Reproduced from [14] with the permission of Elsevier Science Publishers.)

A diagram of the microcolumn cell of the Kratos SF-770 detector is illustrated in Fig. 4.2, where a way of connecting it to a column without intermediate capillaries is demonstrated. Other companies provide interchangeable microbore cells for their spectrometric detectors, for example, Waters, Varian, Micrometrics, Knauer, JASCO, Shimadzu, Gilson, LKB, ISCO, and Perkin–Elmer.

In order to investigate how valid the practice of providing detectors with interchangeable cells, which provide flexibility of use at the expense of being specific for capillary technology, we must establish a relationship between the detector's signal-to-noise ratio and the dimensions of the measuring cell. A photometric detector prcduces a signal that is proportional to the optical density of the solution:

$$\delta = K_1 \varepsilon C \ell = K_1 \log\left(\frac{\Phi_0}{\Phi t}\right) = K_1 \log\left(\frac{I_0}{It}\right), \tag{4.10}$$

where K_1 is the amplifier constant, ε is the molar absorptivity, ℓ is the path length of the cell, C is the solute concentration, Φ_0 and Φt are the fluxes of the incident and transmitted radiation, respectively, and I_0 and It are the corresponding fluxes for the radiation sensor. Since we are interested in very sensitive detection, the measureable transmittance $\Phi_0/\Phi t$ must be close to unity ($\varepsilon C \ell < 0.01$), i.e.,

$$\delta \approx K_1\left(\frac{\Phi_0}{\Phi t} - 1\right) = K_1\left(\frac{I_0}{It} - 1\right). \tag{4.11}$$

The noise in the detector, i.e., the absolute error in the value of the signal, is

$$\Delta\delta = K_1\Delta\left(\frac{\Phi_0}{\Phi_t}\right) = K_1\Delta\left(\frac{I_0}{I_t}\right). \tag{4.12}$$

Since the relative errors of the divisor and dividend are added

$$\Delta\delta \approx \left(\frac{\Delta I_0}{I_0} + \frac{\Delta I_t}{I_t}\right)\delta$$

and because $I_0 \sim I_t$, we have

$$\Delta\delta \approx 2\frac{\Delta I_0}{I_0}\delta. \tag{4.13}$$

We shall only consider the noise associated with the photosensor, for which

$$\Delta I_0 = (a_1 I_0 + a_2)^{\frac{1}{2}} = (a_1 \Phi_0 S_\lambda + a_2)^{\frac{1}{2}}, \tag{4.14}$$

where a_1 is a constant related to the shot noise of the sensor, which is inevitable due to the discreteness of light, a_2 is a constant associated with the dark current in the sensor and thermal noise, and S_λ is the spectral sensitivity of the sensor. If the light flux from the source to the cell is uniformly distributed with a solid angle $\omega = \pi D^2/4x^2$, where D is the diameter of the cell and x is the distance from the chamber to the source, then

$$\Phi_0 = a_3 F_R D^2. \tag{4.15}$$

Here F_R is the intensity of the source in the direction of the chamber, $a_3 = \pi T/4x^2$, and T is the transmittance of the optic system. Given that $a_1 I_0 > a_2$, we have

$$\Delta\delta = 2K_1\varepsilon C\left(\frac{a_1}{a_3 F_R S_\lambda}\right)^{\frac{1}{2}} \frac{\ell}{D}, \tag{4.16}$$

and given $a_2 > a_1 I_0$,

$$\Delta\delta = \left(\frac{2K_1\varepsilon C\sqrt{a_2}}{a_3 F_R S_\lambda}\right)\frac{\ell}{D^2}. \tag{4.17}$$

Equations (4.10), (4.16), and (4.17) demonstrate that the signal-to-noise ratio $\delta/\Delta\delta$ is independent of the cell's optical path length; however, it is a function of the cell's diameter:

$$\delta/\Delta\delta = AD^n, \tag{4.18}$$

where n is between 1 and 2 and depends on the quality of the sensor. This derivation was confirmed by Kok et al. [165], who described a simple modification of the cell in the Waters 440 photometric detector to adapt it for microbore columns. They placed inside the cell a piece of PTFE tubing (internal diameter 0.3 mm, outside diameter 1.8 mm) ground into a cone to fit the shape of the cell. The diameter of the cell, but not its length, was reduced by the modification, and its volume thus decreased to 0.67 μl. Since the path length (1 cm) remained the same, the modification did not affect the concentrational sensitivity of the detector. However, the noise level rose threefold because of the lower light flux through the cell, as predicted by (4.16) and (4.17).

The equations have another important consequence. Although the signal from a photometric detector is independent of the intensity of the light flux, i.e., its radiance, increasing the latter is a good way of depressing the noise. The optical system can also be improved (raising T) in this way by increasing the amount of the source's energy that reaches the sensor. Decreasing a_2 in (4.14) offers yet another way of reducing $\Delta\delta$, e.g., by using a photomultiplier with a cooled cathode as the light sensor.

In general, when developing an absorbance detector for capillary liquid chromatography the instrument maker must strive to keep the cell's optical path length to 1 cm, while any decrease in diameter, which results in an increase in noise, must be compensated for by improving the system's optical performance. A simple substitution of the chamber for a smaller one cannot do this. Thus, the replacement of the cell in commercial photometers must lead either to a decrease in sensitivity (if the length is reduced for constant diameter) or to an increase in the noise level (if the diameter is reduced). Hence it is essential that the absorbance detector be specifically designed for capillary chromatography to ensure the greatest light-gathering capacity.

A successful solution of these problems can be seen in ISCO's μLC-10 absorbance detector [166], which has a cell volume of 0.5 μl and an optical path length of 1 cm. The cell has a threshold detection limit of around 1 pg. This excellent result is achieved by optimizing the detector's optical system and the application of a light storage device made from optically polished rhodium.

Another approach to the miniaturization of the photometer cell is based on a laterally illuminated silica capillary that is joined to the column outlet (e.g., as in JASCO's Familic-100 chromatograph) or forms the lower part of the column itself [98, 99]. This sort of detection is called "on-column" and is the only one that can provide the small detection volumes (down to 0.25 nl) needed when using capillary columns. A diagram of an on-column photometric detector is illustrated in Fig. 4.3.

Photometer capillaries have their own peculiarities. Since a capillary is a cylindrical lens with a short focal length, it strongly affects the light flux passing through it. This nonideal form means that the smallest inaccuracy in installing the capillary causes a deviation of the light beam and a change in the scattering.

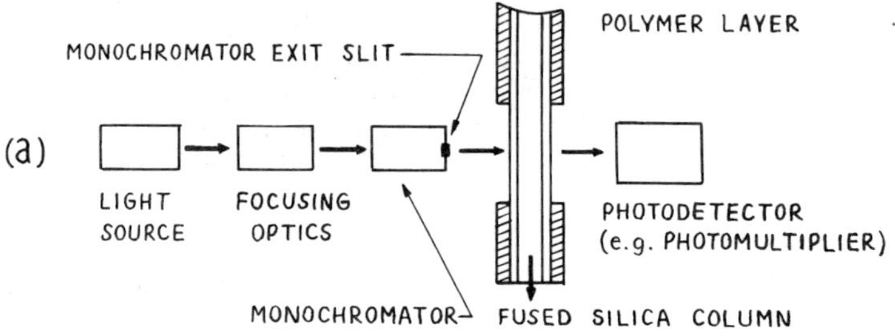

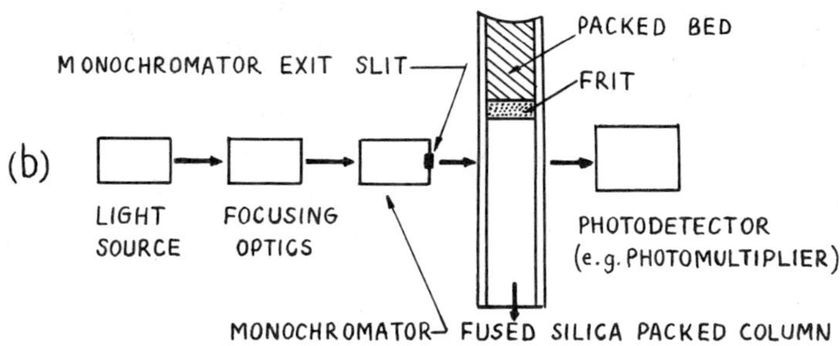

Fig. 4.3. Diagram of an on-column absorbance detector for a) open tubular columns and b) packed microbore columns made from fused silica. (Reproduced from [98] with permission of Dr. Alfred Huthig, Verlag GmbH.)

Using a twin-beam setup, the beam passing through the capillary may be cut off at random during operation, or the light directed to another part of the sensor. Thus, Kuzmin and Mikichur [169] have shown that it is better to use the double-wavelength method for capillary photometers. Two light beams with the same geometry but different wavelengths are passed through the same sample one after the other, the wavelength of one of the beams lying within the range absorbed by the sample, that of the other beam outside it. If the capillary is moved, the scattering changes and the like are approximately the same for each beam.

For on-column detection $\ell = d_c$ and, therefore, the concentrational sensitivity of this sort of detector is very small. The only way of resolving this problem seems to be to reduce the noise level of the detector, as described above, and to increase the signal amplification proportionally.

The first studies using multiwavelength photometry [10, 169] used devices that could not quickly switch the wavelengths. Recently, however, miniature multichannel spectrometers have been built using linear photodiode arrays [167, 168]. These devices aid in the identification and discrimination of the peaks and make it possible for computers to resolve artificially (deconvolute) unseparated compounds, thus substantially increasing the informational significance of microbore chromatographic experiments.

4.3.2. Fluorescence Detectors

There are few detectors that can register a solute in quantities of a picogram, let alone a femtogram. One such detector, however, is the fluorometer. A fluorescing substance only reemits part of the radiation it absorbs, this fraction being defined as the quantum efficiency ϕ. For substances with ϕ close to unity the detection limit using fluorescence is 10^2 to 10^3 times lower than the limit using absorption detection. The fluorescent light flux depends on the concentration of the solute thus

$$\Phi = \phi k_c' k_c'' \Phi_0 (1 - e^{-\varepsilon C \ell}), \tag{4.19}$$

where k_c' is the efficiency of the fluoresce collection, k_c'' is the efficiency of the excitation light collection, Φ_0 is the light flux of the exciting radiation, ε is the molar absorptivity, and ℓ is the length of the optical path of the flow cell. The equation simplifies for dilute solutions ($\varepsilon C \ell \ll 1$) to

$$\Phi f = \phi k_c' k_c'' \Phi_0 \varepsilon C \ell. \tag{4.19a}$$

If the excitation light flux is uniformly distributed within the solid angle $\omega = A_C/x^2$, then the following expression is valid for the emitted light flux:

$$\Phi f = \phi k_c' k_c'' F_R \omega \varepsilon C \ell = \phi k_c' k_c'' F_R \varepsilon C V_d^* / X^2, \tag{4.20}$$

where A_C is the cell's cross-sectional area, X the distance from the source to the cell, F_R the radiation intensity of the source in the direction of the cell, and V_d^* the illuminated cell volume. In contrast to a photometer, a fluorometer measures the absolute value of the light flux, and not the ratio of fluxes. Thus, the signal of a fluorescence detector depends on several factors including the illuminated volume of the measuring cell, the radiance of the light source, the transmittance of the monochromators (filters), and the spectral sensitivity of the photomultiplier. Common fluorometers use mercury lamps, xenon arcs, and deuterium lamps for the light source, depending on the application. Three types of optical system are used to separate out the requisite parts of the irradiated and emitted spectra, viz., the filter–filter, monochromator–monochromator, and monochromator–filter schemes, and each has its own advantages. The design of the measuring cell is very important. An example of a fluorescence detecting cell that successfully overcomes the complexity of problems is the flow cell in the Kratos-970

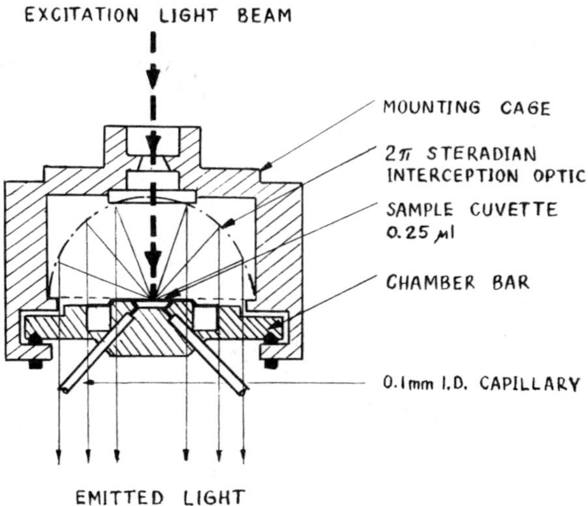

Fig. 4.4. Diagram of the FSA-980 fluorescence detector in the Kratos SF-970. (By permission of Kratos Instruments, New Jersey, USA.)

detector. The version for conventional chromatography has an internal volume of 5 μl (not counting the connecting capillaries), while the microbore version has a volume of 0.25 μl. The cell's design is shown in Fig. 4.4. A concave mirror, with an orifice for the exciting radiation, behind the cell collects nearly all the fluorescence (75% with a cell volume of 5 μl). The limiting detectable concentration for the miniaturized cell ($V_d^* = 0.25$ μl) is about 20-fold lower than that of the standard cell and is $9 \cdot 10^{-9}$ g/ml, which corresponds to 2.2 pg of solute in the cell [164].

Other companies, such as Gilson, Farrand Optical Co., Ikatos Analytical Instruments, and JASCO, produce commercial fluorometers for microbore columns.

Very good sensitivities have been obtained using fluorescence detectors in capillary chromatography for various classes of compounds [12, 101, 129, 170-172]. For example, Koroleva et al. achieved sensitivites better than 10^{-13} mole for DNS amino acids [12, 133]. There has also been progress in constructing fluorometers for capillary columns. The optical and detection layouts for an on-column detector on an open tubular column are shown in Fig. 4.5. The analytical block is mounted on an xyz positioner so that the capillary can be placed at the focus of the exciting light. The detector has a limiting mass sensitivity of around 10 pg, with a linear dynamic range of three to four orders of magnitude. The effective volume of the measuring cell is $\pi d_c^2 \ell^*/4$, where ℓ^* is the length of the illuminated section of the capillary. Given $\ell^* = 2$ mm and $d_c = 60$ μm, we have $V_d^* = 2$ nl.

An analogous setup for on-column detection has been studied, the source being a He–Cd laser with λ = 325 nm [101]. The detection limit for a fused-silica capillary column (d_c = 25 μm) was 1.4 fg with a time constant of 1 sec. Laser fluorometry is thus an extremely promising method for detecting a solute in the effluent from capillary columns. The small radius of a laser beam means that the detecting volume can be reduced to below 1 nl. On the other hand, Eq. (4.19) indicates that the signal of a fluorometer is proportional to the intensity of the exciting radiation. Thus, the very high powers of laser beams means that, in principle, the extreme registration sensitivity of conventional fluorometers may not just be conserved, but increased for capillary columns. Under optimal conditions the power levels of the light flux from a 100-W mercury lamp at 254 nm, a 50-W deuterium lamp at 230 nm, and a 100-W halogen–tungsten lamp at 600 nm are, respectively, 4.5 mW, 0.9 mW, and 17.3 mW [173]. The output power of a laser beam can, however, reach 10 W.

Moreover, even if a laser beam were less intense than that of a lamp, it would still be preferable because the light from any of the lamps is not easily concentrated efficiently on a section of the cell – not a problem for lasers, which emit radiation in one direction only. In addition, laser light is scattered in a plane, unlike light from other sources, which is scattered in every direction. This, together with the coherence and monochromaticity of laser radiation, makes it possible for an investigator to cut out the background radiation due to Rayleigh and Raman scattering by the optical elements and the eluent.

Two types of laser fluorometer cells that may be successfully applied to capillary chromatography are shown in Fig. 4.6. In the drop cell [174] (Fig. 4.6a) the laser light excites luminescence in a drop of effluent that leaves the column and is supported on a stainless steel pedestal which ensures that each drop is the same size and shape. The absence of walls minimizes the scattering from and fluorescence of the optical elements. The greatest difficulty of this design is ensuring the optical integrity of the drops and the reproducibility of their shape. The problem of destroyed bubbles encountered in this method and the impossibility of using gradient elution because of the unstable droplets it produces make the cell shown in Fig. 4.6b [175] more attractive. The eluate flows around the end of an optic fiber placed in the measuring capillary. The stream flows from bottom to top so that bubbles which may be taken to the cell by the mobile phase can escape freely. The optic fiber is placed as close as possible to the focused laser beam so as to increase the fraction of fluorescence that is gathered. Since light is transmitted along an optical fiber by total internal reflection, there is a critical angle above which light entering the fiber is not transmitted. This is used to screen out scattering and fluorescence by the capillary itself. The volume of the cell is deter-

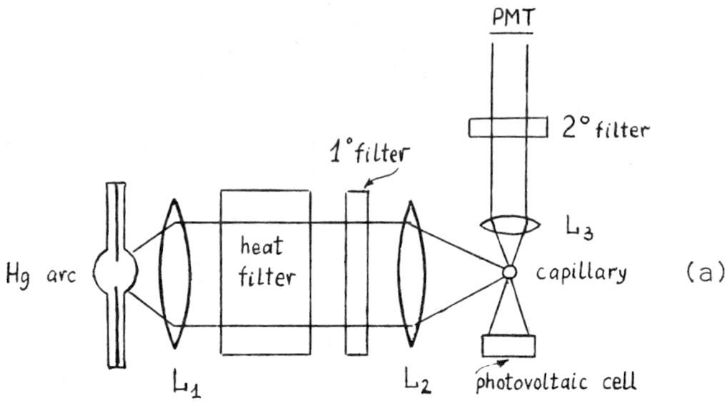

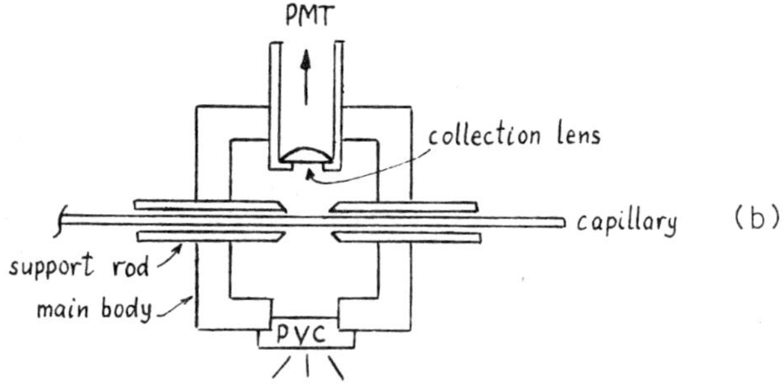

Fig. 4.5. An on-column fluorescence detector for open tubular columns. a) The optical system and b) the detection module. (Reproduced from [100] with the permission of the American Chemical Society.)

mined by the diameter of the optic fiber and can be made sufficiently small if the fiber is placed right at the bottom of a fused-silica capillary.

Detection limits of a few picograms have been obtained from both cell types for determining aflatoxins (λ_{ex} = 325 nm) [174] and antitumor preparations (λ_{ex} = 488 nm and λ_{em} = 590 nm) [175].

Impressive results have been gained with an original hydrodynamic cell in which the chromatographic effluent is injected into the center of a surrounding (sheathing) stream of mobile phase and moves within it without migrating because the flow is laminar. The design for such a fluorometric chamber, which can have a measuring cell volume of 6-150 nl, is shown in Fig. 4.7 [131].

The effluent reaches the cell via tube A and passes along the axis of a cylindrical canal into which the sheathing eluent is delivered via input capillaries (B). A calibrated canal 2 mm long

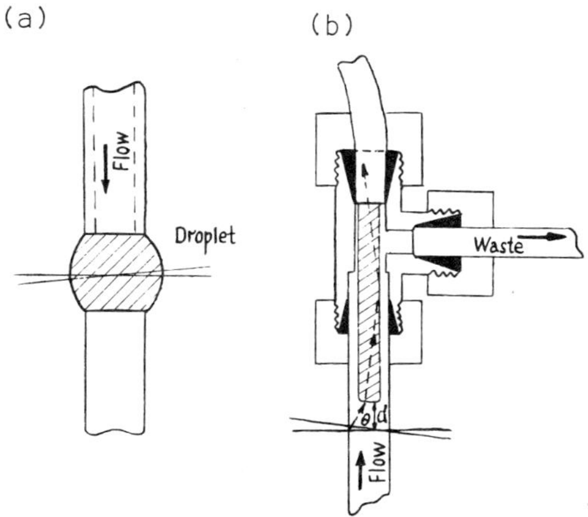

Fig. 4.6. Two flow cells for laser fluorescence detectors: a) wall-less, a drop is formed on a pedestal [174]; b) the cell is a combination of a section of capillary and an optic fiber [175].

and 500 μm in diameter (E) is used to align the flow. In the optical section of the chamber the quartz windows for the exciting radiation and for observing the resultant fluorescence are placed at right angles. The laminar flow prevents the central and sheathing flows from mixing; moreover, it is possible, by changing the speed of the flow, to alter the volume of this wall-less cell; scattering of the exciting radiation is also absent. The cell was tested on the dimethyl ether of mesoporphyrin IX and the disodium salt of protoporphyrin IX with red fluorescence excited by the visible radiation (λ = 488 nm) of Spectra Physics' mark 162 ion-argon laser (8 mW). The laser radiation was focused on the measuring cell by a spherical lens with a focal length of 10 cm, while the fluorescence was collected by a microscope objective and directed onto a photomultiplier through focusing optics. Filters with cutoff wavelengths of 520 and 625 nm were employed to eliminate the scattered exciting radiations. The detection limit was 53 pg for a signal three times the noise level.

The sheath flow principle underlying this laser fluorometer significantly reduced the stray light, avoided the adsorption of the sample on the windows, and most importantly provided the small effective volume of the cell for registering the eluate needed in capillary chromatography. We must also point out that only a relatively low-powered visible laser was needed to observe fluorescence in this cell.

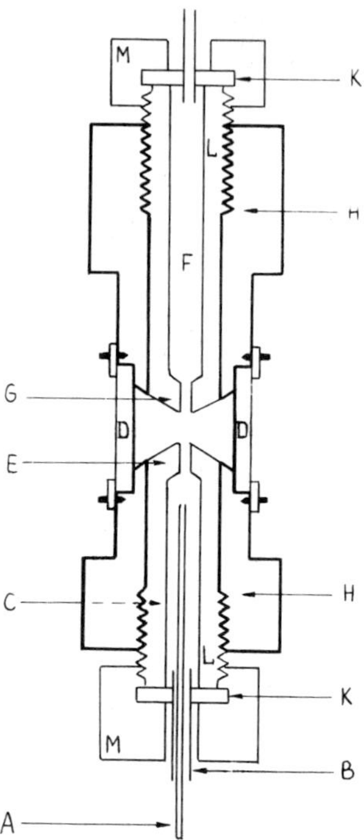

Fig. 4.7. Hydrodynamic cell for a laser fluorescence detector. A) Inlet tube for effluent; B) inlet tube for sheathing flow; C) inlet canal of sheathing flow; D) quartz window; E) inlet of forming capillary (500 μm in diameter); F) exit canal; G) exit forming canal (500 μm ID); H) cell body; K) sealing washer; L) inlet and exit probes; M) steel nut. (Reproduced from [131] with the permission of the American Chemical Society.)

Even in the above studies the potential of laser fluorometry had not been fully realized. Krypton fluoride lasers can be used for UV excitation, making the detection limit fantastically low, i.e., a few femtograms [129, 130].

Laser fluorometry has other features favorable to capillary chromatography. The great intensity of laser sources makes it possible to implement double-photon excitation, which is equivalent to excitation by radiation with half the actual wavelength of the laser, i.e., $\lambda = \lambda_L/2$. This will help filter out the scattered exciting radiation from the fluorescence. Since lasers are the only source of nanosecond impulses of light, they might be used to produce post-illumination luminescence free of the exciting radiation. And, finally, the polarization of laser radiation means scattered light can be eliminated by using polarizing filters.

4.3.3. Refractometers

Of the three types of refractometers now produced industrially, i.e., Fresnel, interferometer, and hollow prism (deflection type), the detector with the smallest volume is the Fresnel refractometer (made by Laboratory Data Control), and the one with the greatest sensitivity is the interference RI detector (Optilab). A detector for capillary chromatography that uses both interference and polarization is described by Aleksandrov's group [11, 176]. It requires a 1-mW LG-56 helium—neon laser (wavelength 630 nm). The beam from the source is split into two beams with equal intensities and polarized in orthogonal planes. These are passed through the measuring and reference cells, respectively, and then united into a single beam by a crystal of Iceland spar. The resultant wave is elliptically polarized with any change in the refractive index of the substance in the measuring cell changing its polarization. The phase difference between the components of the elliptical polarizations can then be measured. It is given by

$$\phi = (2\pi/\lambda)(n_2 - n_1)\ell, \qquad (4.21)$$

where λ is the wavelength, n_1 and n_2 are the refractive indices of the media being compared, and ℓ is the length of the optical path.

The phase difference is measured using a Senarmon compensator with an electrooptical modulator.

Although in principle it should be possible to measure the phase difference to one second of arc and, hence, changes in the refractive index to $5 \cdot 10^{-10}$ for $\ell = 0.1$ cm, in practice this sensitivity is difficult to achieve because of the difficulties involved in aligning the temperatures of the two cells and stabilizing the path differences caused by the optical elements. By using an effective heat-exchanger it is possible to measure a refractive index difference of 10^{-8} RI for a detection volume of 1 μl and a difference of 10^{-7} RI for a volume of 0.1 μl. A capillary with an internal diameter of 0.15 mm and volume about 2 μl was used as the heat exchanger, though obviously this increases the overall dynamic volume of the system. However, this is not catastrophic [see (3.55)] because the volumetric standard deviation σ_V for the spreading in a capillary may, according to Taylor and Golay's equation [151, 152], be considerably less than the total volume of the capillary, and Atwood and Golay [177] have shown that for short tubes it is even smaller. Apparently, the compromise between the necessity of keeping the temperature in the cell constant and reducing the extra-column dead volume may be found by using the zig-zag capillaries suggested by Katz and Scott [178] for the heat exchange.

The device was designed with automatic zeroing, an internal sensitivity calibration control, and wall that could be maintained at a temperature between 15°C and 40°C. The detection limit (for albumin) was 0.15-0.2 ng.

This detector is used with microbore columns in the KhZh-1309 liquid chromatograph (Table 4.4), which is produced in the USSR, and which has been successfully applied to determine the molecular mass distribution of polymers [176] by gel permeation chromatography

4.3.4. IR Spectrometers

Infrared absorbance detection is often much better than UV photometry. Though less sensitive, it is more flexible. For example, if we tune the detector to look for C—H bond stretching, λ = 3.5 μm, then we have a universal detector. However, if we tune it for a wavelength of 5.8 μm, then we are selectively looking for compounds with C=O groups. The basic difficulty is to match the necessary transparency of the solvent in the appropriate part of the spectrum with its requisite eluting strength. Both the ordinary dispersion spectrophotometers and the newly developed Fourier-transform IR (FTIR) spectroscopes [179] can be used to detect the eluate. The Fourier devices are better because they combine submicrogram sensitivity (q_{min} < 1 μg), the ability to monitor more than one wavelength at a time, and the ability to provide full IR spectra at any point in the chromatograph. However, the device is rather expensive, with a current cost of around £20,000.

There is a great deal in the literature about connecting microbore columns on line to dispersion and Fourier-transform IR spectrometers [124-128, 180-182, 184]. One way of solving the problem of the absorption of infrared light by the mobile phase is to collect the effluent on an alkali-metal halide and then evaporate off the solvent. Since this procedure is difficult to automate, the collected fractions are usually analyzed off-line. Obviously, this method limits the number of compounds that can be analyzed to those that are less volatile than the solvent. Since the peak volumes in conventional high-performance liquid chromatography may be several milliliters in volume, the effluent must be concentrated before being taken to the metal halide powder (KBr or KCl) [183]. This stage makes the process even more difficult to automate. However, capillary chromatography is much more suitable for this purpose. Jinno and Fujimoto [181, 184] described the use of a crystalline plate of KBr to interface a microbore column and a Fourier-transform IR spectrometer. The eluate, flowing at 5 μl/min, is collected continuously on the crystalline plate, which is then automatically transferred to the optical section of the spectrometer. The start and end of the collection of the peak are determined by a UV detector. Jinno and Fujimoto were able to automate the process because the volumes of the peaks were small enough for the whole chromatogram of a multicomponent mixture to be deposited on the crystalline plate.

It later became possible to miniaturize the system of connecting an FTIR spectrometer off-line with a microbore column. The interface involves a computerized carousel with KBr-filled cups that are used to collect the fractions [183]. Although these interfaces reduce the constraints on the solvents used, they have been overtaken in terms of simplicity, convenience, and universality by FTIR detection in a flow cell connected on-line to a chromatograph.

If the mobile phase is only moderately absorptive in the infrared region, then a flow cell made of sodium chloride 0.5 μl in volume, or one made of potassium bromide 1.2 μl in volume, is sufficient for connecting a commercial Fourier-type spectrometer on-line to a microbore column [124]. The interferogram of the effluent is then stored on disk, and a stacked plot of absorption versus elution volume can be constructed later. The detection limit for diethyl phthalate was 1 μg.

A flow cell made from a flattened section of PTFE tubing and providing an optical path length of 30 μm has been used [125] for the detection of hydrocarbons, alcohols, and vitamins from their CH, OH, and CO groups, the limiting sensitivity being 0.5 μg. The solvents used were perdeuterium and perhalogen compounds, such as N-trifluoromethylperfluoromorpholine.

The work on FTIR on-line detection [124, 126, 128, 180] has revealed a series of advantages for capillary liquid chromatography. Primarily, this sort of chromatography creates conditions conducive to detection because of the significantly smaller sample dilution in the column. This means that the length of the optical path in the detector can be brought down to 1 mm without any loss in sensitivity and ensures that many solvents are tolerably transparent. Second, the small solvent consumption that is typical of capillary liquid chromatography means that it is still economical to employ exotic, but expensive, solvents that are transparent in the infrared region. Finally, the combination of a superefficient system of microbore columns, containing millions of theoretical plates [16, 52], and FTIR detectors opens up promising avenues for the identification of components from very complicated mixtures. The FTIR spectrum of indole is given in Fig. 4.8a. The spectrum was taken from the maximum of a chromatographic peak during the separation of a test mixture made up of nine amines. The separation was carried out on a 1 m × 1 mm microbore column filled with a polar amino cyano-grafted phase (Partisil-10-PAC, Whatman) [126]. A comparison spectrum of indole in the same eluent ($CDCl_3$) is given in Fig. 4.8b.

These spectra clearly show the speed and unambiguity of identification of the peaks with an FTIR detector. On-line FTIR detection has become very widely used with steric-exclusion chromatography [124] because it is relatively easy here to choose a single-component solvent with wide transmission windows.

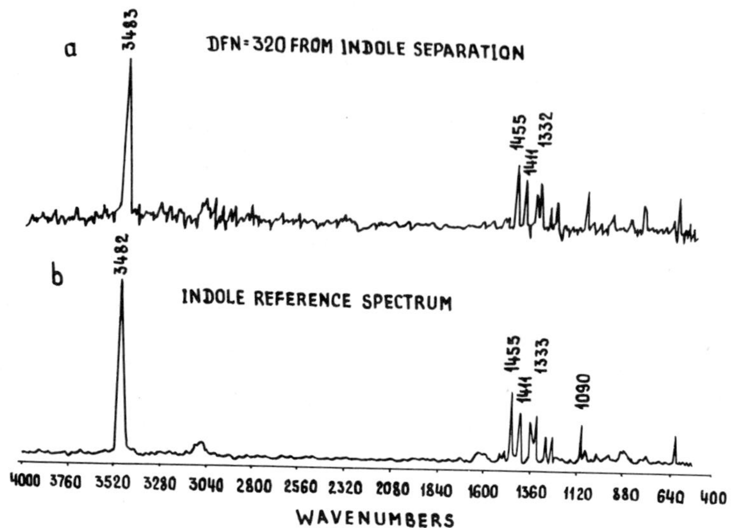

Fig. 4.8. a) FTIR spectrum of indole at the maximum of a chromatographic peak during the separation of a test mixture of amines on a 1 m × 1 mm microbore column filled with Partisil-10-PAC. The FTIR detector was connected on-line, with a KBr flow cell (V_d = 3.2 μl, ℓ = 0.2 mm). The mobile phase was $CDCl_3$ and the amount of solute in the peak 6 μg. b) Comparison FTIR spectrum of indole. (Reproduced from [126] with the permission of the American Chemical Society.)

However, solvents such as Freon-113, carbon tetrachloride, $CDCl_3$, and chloroform even retain sufficient transparency between 400 and 4000 cm^{-1} when moderate amounts of polar modifiers, e.g., acetonitrile, are added. Thus, it is possible to also carry out normal-phase liquid chromatography with these binary solvents in an isocratic regime [128].

A serious defect with FTIR detection concerns the difficulties encountered in applying gradient elution and reversed-phase separation systems. This can be overcome to a great extent by the use of superefficient microbore steric-exclusion chromatographs. In spite of the inherently low selectivity in this technique, a column with 10^5 to 10^6 theoretical plates can be used to separate by size molecules with very close chemical compositions (see Fig. 2.1). This, it seems to us, is the path along which FTIR detection in microbore columns should advance.

4.3.5. Electrochemical Detectors

The classification of this type of detector is disputed. Some authors (e.g., White [162]) have used this category to include voltammetric and potentiometric devices together with those that measure dielectric permittivity or conductivity; i.e., any system using an electrical property for detecting the presence of a substance in an effluent is thus classified. Others consider that the term "electrochemical detection" may only be used when the electrolysis of electrically active species is involved, i.e., voltammetry [186].

The latter point of view is more acceptable, and we shall adopt the following terminology. Depending on how complete the electrolysis, we distinguish between coulometric (100% electrolysis) and amperometric (partial electrolysis) detection. The electrode potential during amperometric detection is not sufficient to oxidize or reduce the solute completely, as happens during coulometric detection.

Current detection is considered to be the more sensitive because as the magnitude of the potential is increased, the electrode becomes more contaminated and the background current rises.

The half-wave potential for a given electrode is characteristic of the species under consideration, and hence electrochemical detection can be extremely selective, especially when using two-electrode systems in various configurations. The detection limit of an electrochemical system is comparable with that of fluorometry, reaching a lower limit of 50 fg [185].

The most useful aspect of electrochemical detection is the wide variety of approaches that are possible, such as two-electrode systems, coulostatic detection, potential scanning, and potential or charge modulation (pulse modulation). This makes the technology very flexible and makes it possible to design the very small flow cells needed for capillary liquid chromatography without sacrificing sensitivity.

In most cases the measuring cell includes three electrodes, i.e., the working, reference, and auxiliary electrodes. The reference electrode can be excluded, but then the absolute value of the potential at the working electrode becomes indeterminate and the ability to compensate for deviation of the potential from the prescribed value is lost. This may show itself as a nonlinear response from the detector.

The two most common flow cell configurations for amperometric detectors are the thin-layer and wall jet configurations.

The Faraday current I through a thin-layer cell depends on the volumetric flow rate of the liquid phase, F, and on the dimensions of the cell [187, 188]:

$$I = n_e \Phi C F [1 - \exp(-Sh \cdot D_m \ell b / 2Fd)], \qquad (4.22)$$

where Φ is Faraday's constant; C is the concentration; n_e is the number of electrons involved in the redox reaction; Sh is Sherwood's number; ℓ, b, and d are the length, width, and height of the measuring canal of the cell, respectively; and D_m is the diffusivity.

At small eluent rates and given that $D_m \ell b/Fd > 0.334$, the current ceases to depend on the flow rate [187, 188], i.e.,

$$I = n_e \Phi C F\left(0.267 + \frac{D_m \ell b}{Fd}\right). \tag{4.23}$$

The effective volume of the cell is determined by the dimensions of the orifice in the film spacer and by the thickness of the spacer. When this sort of cell is miniaturized, the reduction in the length and width results in a fall in the concentrational sensitivity. The sensitivity of a cell 0.15 μl in volume, for example, is only 3% that of a conventional thin-layer cell 2 μl in volume [189]. However, miniaturization significantly cuts down the noise level, and so the detection limit in terms of mass of substance is only raised tenfold. We shall demonstrate later that the special abilities of electrochemical detectors, such as parallel-opposed double-electrode detection, enable the cell designer to compensate, indeed more than compensate, for the loss of sensitivity.

As we shall show below, the effective volume of a detector with a wall jet cell depends on the surface area of the working electrode and the distance between it and the exit of the column. Both sorts of flow cell are easy to miniaturize, and measuring cells suitable for packed microbore columns with volumes of 0.3 μl [190] and 0.15 μl [189] have been described. Moreover, Šlais and Kouřilova [191, 192] have managed to obtain a measuring cell around 20 nl in volume, and Šlais and Krejči [93] a cell less than 1 nl in volume. From the data in Tables 4.1-4.3 it can be seen that these correspond to the most severe requirements on the cells for ultramicrobore packed and open tubular columns.

The measuring volumes of some commercial electrochemical detectors are within the boundaries set by capillary liquid chromatography, e.g., Pye Unicam's PU 4022, which has a cell volume of 0.5 μl, and the LKB 2143, which has a volume of 1.5 μl. The main difficulty with these detectors is connecting them to the microbore columns with the least volume of the communicating capillaries. One way of doing this is shown in Fig. 4.9 [193]. The inlet orifice of a TL-5 thin-layer cell (made by Bioanalytical Systems) was reamed out to the outside diameter of the microbore column (1.6 mm), and the column was fitted directly into the cell and sealed with a section of flared PTFE tubing. In addition to reducing the thickness of the PTFE spacer to 0.05 mm this enabled the total volume of the thin-layer canal to be reduced to 0.9 μl.

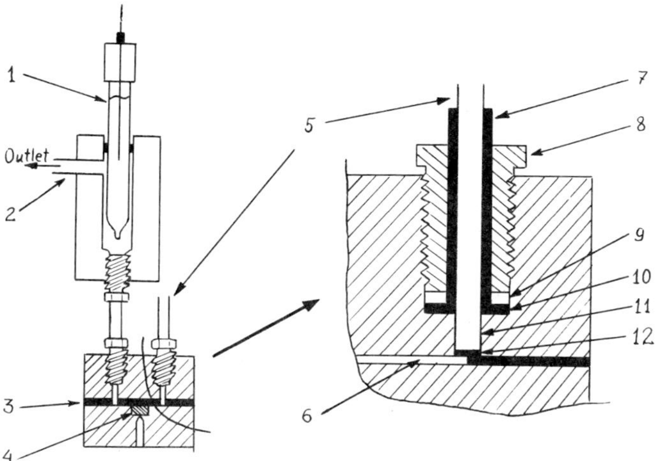

Fig. 4.9. Modification of Bioanalytical System's TL-5 thin-layer cell for detecting effluent from a microbore column. 1) Reference electrode; 2) auxiliary electrode; 3) PTFE spacer; 4) working electrode; 5) microbore column; 6) thin-film canal; 7) PTFE hose; 8) polypropylene bush; 9) steel washer; 10) flared end of PTFE tube; 11) canal reamed out to 1.6 mm; 12) steel frit. (Reproduced from [193] with the permission of the American Chemical Society.)

The choice of material for the reference (Ag, AgCl) and auxiliary electrodes was no problem, it being possible to utilize for the latter the steel capillary used to transfer the effluent from the column to the cell. However, the material for the working electrode has to meet stringent requirements. Glassy carbon is highly recommended. It can be polished to a very smooth surface, which means that very thin spacers can be used inside the cell. The operating potential region of the electrode is from −1.3 V to +1.5 V, which is wide enough for registering a variety of compounds.

A miniaturized thin-layer cell (volume 0.3 μl) is illustrated in Fig. 4.10; the working electrode is made from glassy carbon and is 3 mm in diameter. Hirata et al. [189] report a thin-layer cell 0.15 μl in volume using pressure-annealed pyrolytic graphite. This turns out to be a very good electrode material for miniaturized electrochemical cells because of its ideally flat surface, which is resistant to deformation. In order to get reproducible detection the surface of the electrode is anodically oxidized.

Platinum electrodes have been successfully applied in capillary chromatography. Figure 4.11 contains a diagram of a wall jet

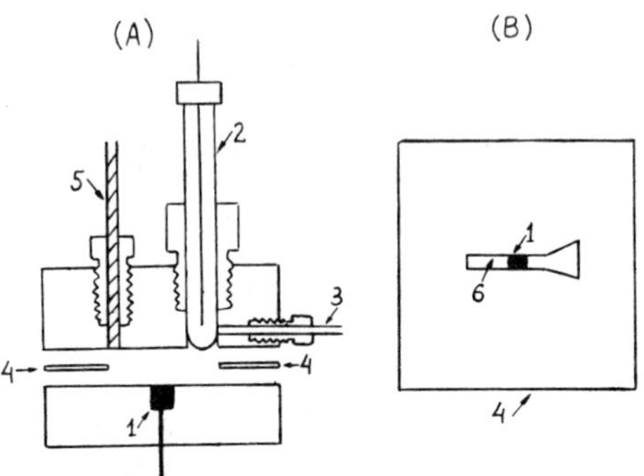

Fig. 4.10. Design of thin-layer cell for an electrochemical detector and its connection to a microbore column. A) View of the assembly; B) top view of the spacer. 1) Working electrode made from glassy carbon; 2) reference electrode (Ag/AgCl); 3) auxiliary electrode (steel capillary); 4) spacer (PTFE sheet); 5) microbore column; 6) orifice in spacer. (Reproduced from [190] with the permission of Elsevier Science Publishers.)

cell in which the measuring electrode is a platinum filament and the auxiliary electrode is a steel capillary placed coaxially around the wire. The cell's dynamic volume is 1 nl. Later the cell's design was simplified (auxiliary electrode removed), but the cell's volume was increased to 20 nl [121].

Very good sensitivities have been achieved using platinum working electrodes in three-electrode [93] (q_{min} = 0.05 pg) and two-electrode [121, 122] (q_{min} = 3 pg) cells. Curiously, the two-electrode cell provided a signal that was linearly dependent on the concentration, and for eluent rates between 0.1 and 0.5 μl/sec the device worked as a concentration- and not a mass-sensitive detector. However, the advantages of a platinum electrode must be set against the possible long-term oxidation of the metal's surface.

Knecht et al. [102] obtained the best results with regard to the electrochemical detection of solute eluted from capillary columns. They used an on-column detection system with the indicator electrode being a single graphite fiber 9 μm in diameter. The fiber electrode was placed about 0.7 mm inside an open tubular capillary column (d_c = 15 μm) using manipulators and a microscope. The combined reference and auxiliary electrode was made of silver wire coated with silver chloride and placed near the exit to the column. A glass container filled with 0.1 M potassium chloride was glued (using an epoxy resin) around both electrodes. The system worked

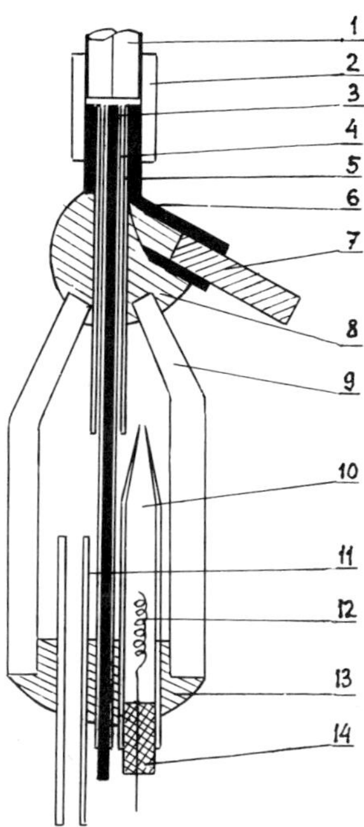

Fig. 4.11. Design of filament cell for an electrochemical detector for capillary columns. 1) Capillary column; 2) PTFE tube; 3) platinum wire; 4, 5) glass capillaries; 6) steel capillary; 7) inlet tube for auxiliary electrode; 8, 13) epoxy resin; 9) glass vessel with reference electrode; 10) reference electrode; 11) exit capillary; 12) silver wire; 14) rubber tubing. (Reproduced from [93] with the permission of Elsevier Science Publishers.)

for eluent rates less than or equal to 1 nl/sec in a coulometric regime with an electrolysis efficiency for catechol and methylcatechol of 90-100%. The detection limit for these substances was around 1 fmole. Although Knecht et al. did not evaluate the effective hydrodynamic volume of the measuring cell, given the geometric dimensions it must have been significantly less than 1 pl.

The systems we have described are appropriate for potentials between −0.4 and +1.2 V with respect to the reference electrode. They are thus undoubtedly useful for easily reducible or easily oxidizable species. In order to widen the applications of electrochemical detectors to less-reducible species, the mercury drop electrode, which can function reliably down to −2 V, must be miniaturized. Hanekamp et al. [194] have described such a detector, and it

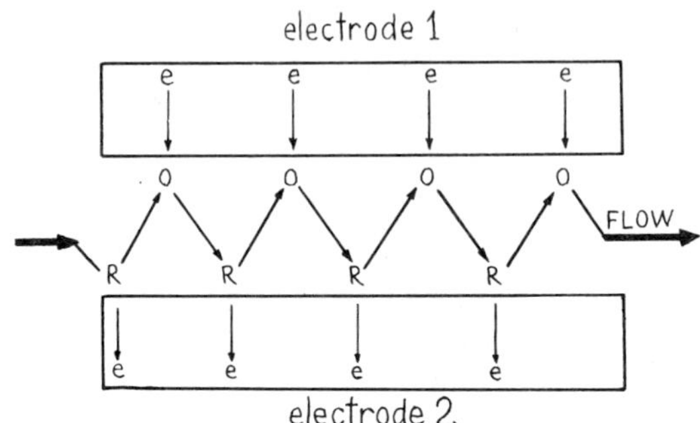

Fig. 4.12. Design of an electrochemical detector with two parallel working electrodes. Electrode 1 is given the anodic potential and electrode 2 the cathodic potential. The anode and cathode currents are both increased by each conversion.

was later shown that the dispersion in its flow cell is less than that in a UV-detector cell 0.7 μl in volume [165].

Because detection at potentials greater than +1 V or lower than −1 V involves a considerable loss of selectivity and sensitivity, Blank [195] has suggested using two working electrodes in series for species with large redox potentials. The species being detected is electrolzyed at the upstream, generator electrode using high absolute potentials. Then the products of the electrolysis are registered at the downstream, detector electrode using smaller potentials. Goto et al. [196] have described a miniaturized version of the thin-layer device with two series working electrodes that is appropriate for species which are electrochemically reversible (e.g., the catecholamines). The cell's design only differs from that shown in Fig. 4.10 by the presence of the extra electrode and the dimensions of the orifice in the PTFE film spacer. The anode and cathode are supplied with different potentials, +0.8 V and −0.2 V, respectively, and the cathode is placed downstream of the anode. The reduced variant of the reversible species is oxidized at the anode and the product of the reaction then reduced at the cathode. Thus, the cathode is only responsive to reversible redox couples.

The parallel-opposed configuration, in which two working electrodes are placed in parallel and close together in a thin-layer cell, has a special value for capillary chromatography (Fig. 4.12) [186, 197]. The electrodes are placed as close together as possible and are supplied with the anodic and cathodic potentials needed for the reversible reaction between the components of the redox pair being analyzed. As illustrated in Fig. 4.12, each conversion is accompanied by an increase in the current generated by each of the electrodes. The current amplification depends on the number of conversions; i.e., it is raised by slowing the effluent flow or diminishing the thickness of the cell [199, 132], viz.,

$$\Phi_e = \frac{[I_a]}{I_s} = -0.07 + \frac{D_m \ell b}{dF}, \tag{4.24}$$

where Φ_e is the current amplification, I_a and I_s are the anodic and coulometric currents, respectively, and ℓ, b, and d are the length, width, and height of the cell.

Goto et al. [132] have developed a miniaturized parallel-opposed thin-layer cell with a volume of 1.2 μl. It yields current amplifications of as much as Φ_e = 19.5 for effluent rates of F = 1.4 μl/min, which is in agreement with Weber and Purdy's theory [199]. This does not mean that the effective current amplification might not be raised by enhancing the polish on the electrode surfaces or by using very thin spacers (less than 5 μm). Usually, the current signal from an electrochemical detector is reduced both by smaller cell dimensions (b and ℓ) and by slower flow rates. Thus, the increase in the signal yielded by a parallel-opposed electrode system means that the basic dilemma of capillary chromatography, i.e., reducing the cell dimensions and flow rate while conserving the mass or concentration sensitivity, can be overcome.

The signal from both electrodes may be recorded separately, cumulatively, or differentially, thus significantly expanding the range of detection. The system described above has been successfully applied to the superefficient detection of catecholamines [198] and biogenic amines [201].

Electrochemical detectors are also promising both for the fast variant of capillary chromatography [200] and for the ultrasensitive variant, the increased sensitivity of the detector at high flow rates then becoming a virtue. At the same time, the inertia of the detector is not very large.

The flaws of electrochemical detectors are the destruction of the detected species, the difficulties of implementing gradient elution, and the presence of a background signal when using a reduction regime due to traces of oxygen and reducible metal ions in the eluent.

4.3.6. Electrical Conductivity Detectors

Species that are fairly strong electrolytes (salts, bases, or acids) can be detected with a tolerable sensitivity by measuring the electrical conductivity of the solution. Commercial detectors

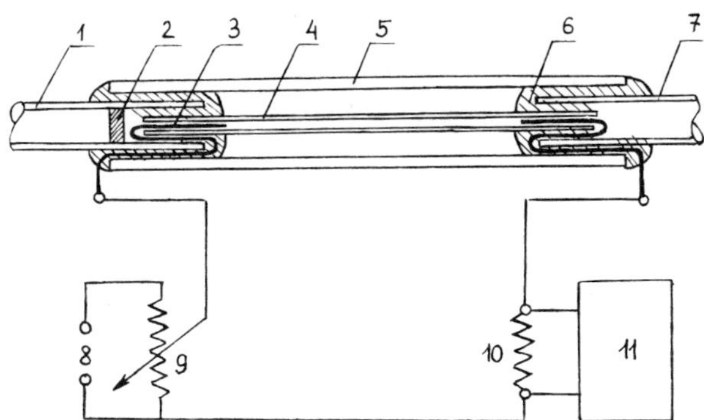

Fig. 4.13. Diagram of a conductivity detector for microbore liquid chromatographs. 1) Packed glass microbore column; 2) glass-wool filter; 3) platinum wire (0.1 mm in diameter); 4) glass capillary (0.15 mm ID); 5) glass capillary (1.3 mm ID); 6) epoxy resin; 7) outlet glass capillary; 8) 30-V battery; 9) 10-kΩ potentiometer; 10) 100-kΩ resistor; 11) recording millivoltmeter. (Reproduced from [203] with the permission of the Czechoslovakian Academy of Sciences.)

of this type usually have cells 2-5 μl in volume with two platinum electrodes and yield concentrational detection limits of 10^{-6} g/ml in moderately conducting eluents. The less conductive the mobile phase is, the greater the sensitivity of the analysis this sort of detector can provide. Unfortunately, ion-exchange, ion-pair, and reversed-phase suppressed-ionization chromatographies all require electrolytes in the eluent which have a deleterious effect on the sensitivity of conductivity detectors.

This sort of detector was therefore not widely applied until the discovery of a method of reducing the background electrical conductivity. The method involves the inclusion of what are called "suppressor columns" downstream of the analytical ion-exchange column. These columns contain H^+ or OH^- ion-exchange resins and substantially reduce the electrical conductivity of the mobile phase. Suppressors laid the foundation for a very elegant analysis method, ion chromatography. It would have been inappropriate here to discuss the rarely used conductivity detector had not both open tubular and packed capillary ion-exchange columns [42-44, 81, 90] and, more importantly, suppressors appropriate to the scale of capillary technology [202] been developed on the basis of hollow ion-exchange fibers. The very simplicity of the cell lends itself to miniaturization. A Czechoslovakian design of a detector cell 0.1 μl in volume is shown in Fig. 4.13 [203]. The cell is affixed directly to the end of a packed glass microbore column (d_c = 0.5 mm) and sealed with epoxy resin. The concentrational detection limit (for acetic acid) is $7 \cdot 10^{-7}$ mole/liter, and the linear range extends up to $2 \cdot 10^{-4}$ mole/liter. It is worth noting that conductivity detectors, when utilized with ion chromatography, can compete with indirect UV photometry [204]. In the latter a UV absorber is added to the eluent, and so the presence of compounds that do not absorb ultraviolet light cause inverted peaks. This method, too, has been applied to capillary liquid chromatography [205].

4.3.7. Potentiometric Detectors

Since ion-selective electrodes have been miniaturized, the potentiometric technique has become attractive for the capillary chromatography of ions when a very selective detector is required. If an ion-selective electrode is made from an element in equilibrium with its ions in solution, then the potential of the electrode will be related to the activity and hence the concentration of the ions via the Nernst equation.

In order to transform this potential into a voltage a reference electrode must be introduced. The ion-selective electrode can be made from glass membrane, an ionic crystal, an adsorbed layer of a compound with ionogenic groups, or a membrane made from a liquid ion-exchanger. The speed with which a membrane electrode can respond to concentration changes depends on the thickness of the membrane and its permeability with respect to the ion. The inertia of an ion-selective electrode is, however, well within the restrictions set by capillary liquid chromatography. Time constants of 10 msec [206] and 7 msec [207] have been achieved for membrane microelectrodes.

It should be noted that measuring a difference between the potentials of the ion-selective and reference electrodes presents no serious difficulties. The relevant apparatus is similar to a good pH-meter. Manz and Simon [103] have produced a tiny potentiometric cell using a membrane ion-selective electrode that is about 1 μm in diameter at the tip. The electrode is placed directly into an open tubular capillary column. It is shown in Fig. 4.14. The membrane is a liquid cationite, a 3% solution of potassium tetra-p-chlorophenylborate in 2,3-dimethylnitrobenzene. The reference electrode (Ag—AgCl) was placed in a micropipette filled with KCl solution. Manz and Simon also proved that the effective volume of the cell lay between $5 \cdot 10^{-4}$ and 10^{-5} nl for an analytical column with d_C = 25 μm. The mass detection limit for K^+ ions against a background of NH_4^+ ions was $5 \cdot 10^{-13}$ mole, even though the membrane was not very selective for these two ions.

Manz and Simon clarified details of the detector construction, such as the electrode support, later [104]. Using on-column detection with an anion-selective electrode an effective cell volume on

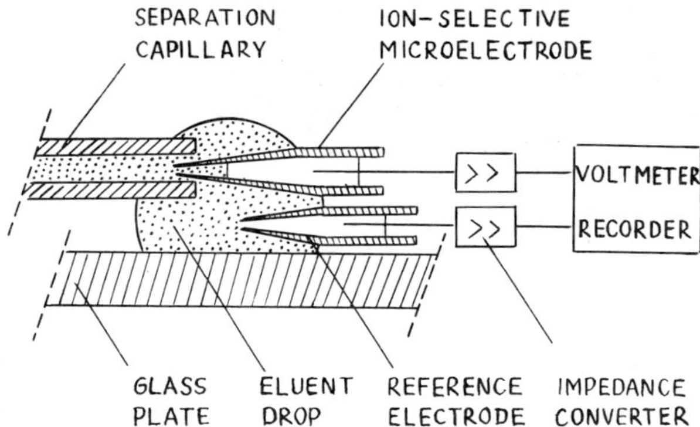

Fig. 4.14. An on-column potentiometric detector for open tubular capillary columns. (Reproduced from [103] with the permission of the American Chemical Society.)

the order of a femtoliter was achieved. The electrode was placed inside the column using a microscope and manipulators, in the same way as the fiber electrode in the previous section was inserted.

4.3.8. Gas Chromatography Detectors

Since gas chromatography has been a routine analytical tool for such a long time, very sensitive and selective detectors have been developed. This has stimulated interest in connecting liquid chromatographs to gas chromatographic detectors. As in all applications of mass spectrometers for high-performance liquid chromatography, there are two approaches: the effluent may be evaporated directly inside the detector or it can be transported there by a moving band or wire through an evaporation oven. The transport systems are described in detail by Scott [159]. They have not proved successful for collecting effluent either for conventional or capillary high-performance liquid chromatography [208, 209], mainly because of transporter wobbles, tears in the band, the band brushing against the nebulizer, selective evaporation of the solvent, and the gradual contamination of the transporter. The direct evaporation of the solvent in a gas chromatography detector is hampered by the significant volume of vapor produced when the flow rates typical for high-performance liquid chromatography (1.0 ml/min) are used. Several research groups have therefore been directing their attention to connecting gas chromatography detectors to capillary liquid columns, and thus utilize the low effluent flow rates.

Gas chromatography detectors can be classified as follows: 1) flame ionization detectors (FID), which are applicable to most organic compounds; 2) flame photometer detectors (FPD), which are selective for sulfur and phosphorus compounds; 3) electron capture

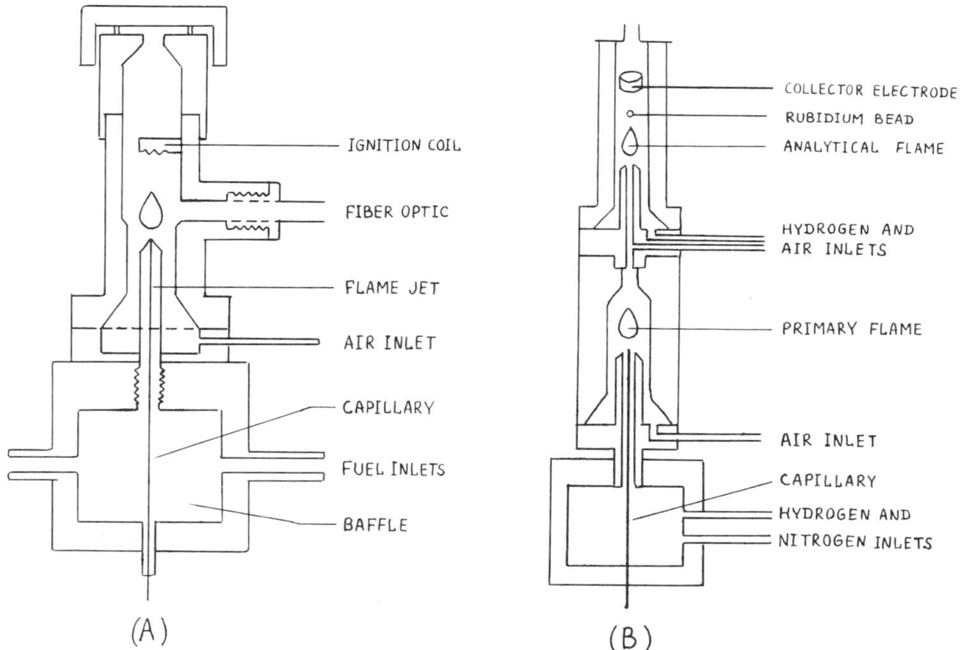

Fig. 4.15. Flame detectors for microbore liquid chromatographs. A) Flame photometer (reproduced from [72] with the permission of the American Chemical Society). B) Twin-flame thermionic detector (reproduced from [212] with the permission of Elsevier Science Publishers).

detectors (ECD), which are specific for halogen- and nitrogen-containing compounds; and 4) thermionic detectors (TID), which are very sensitive, especially to phosphorus- and nitrogen-containing compounds.

The application, without modification, of commercial electron capture devices to microbore chromatographs is described in [210, 211]. However, the results are not impressive. A great deal of success has been achieved with flame detectors [70-72, 143, 212]. It must be noted that in order for the effluent from a microbore column to be efficiently introduced into a flame detector, its atomization and the evaporation of the solvent must be optimized. This stage is common to other detector types, e.g., evaporation analyzers [235] and mass spectrometers with direct liquid injection [116]. McGuffin and Novotny [71, 72, 212] have constructed flame detectors in which the efficiency of the dissociation is easily regulated by changing the flow rates of the fuel and atomizing gases.

The threshold detections for a flame ionizer [70, 71] and a flame photometer [71, 72] are around 1 ng/sec, while for a dual-flame thermionic detector [212] it is two orders of magnitude lower. This yields a good signal-to-noise ratio for detection of 1 ng, given a time constant of 10 sec.

Designs for a flame photometer and a thermionic detector are given in Fig. 4.15. In a flame photometer (Fig. 4.15a) all of the effluent is fed via a concentric nebulizer into a cool diffusion flame that is rich in hydrogen. The characteristic optical emission of the sample is then measured.

The fuel gas (hydrogen) and an incombustible nebulizing gas (nitrogen) are mixed in a chamber. The burner has porous discs so that the air can diffuse evenly into the hydrogen-rich flame. The steel nozzle for the flame has an internal diameter of 0.76 mm, while the glass capillary used to feed the effluent into the flame is 10 cm long (internal diameter 50 µm and external diameter 0.6 mm). One end of the capillary is attached to the microbore column with a section of fluoroplastic tubing, while the other end is set into the nozzle so that it juts 1 mm into the center. The aluminum combustion chamber allows the combustion products to escape freely without losing any of the light. An optical fiber sensor is fixed into the wall of the combustion chamber. This picks up light from the cooler core of the diffusion flame and passes up to 25% of the light to the photomultiplier via an interference filter for 530 nm. The amplified signal from the photomultiplier is filtered with a time constant of between 1 and 30 sec and recorded. The detector's reliability depends on the efficiency of the nebulizer and requires a certain rate of solvent evaporation. This means that the performance depends on the flow rate of the nebulizing gas; the viscosity, thermal capacity, surface tension, and flow rate of the solvent; the temperature of the flame; and the initial size of the droplets. The optimum deliveries are hydrogen 55 ml/min, nitrogen 90 ml/min, and air 75 ml/min. The introduction of an organic solvent into the eluent changes the optimum ratio of fuel to oxidant, and hence a flame photometer cannot be recommended for gradient elution.

It has been observed that the addition of acetonitrile to the aqueous eluent starts quenching the emission of phosphorus at concentrations as small as 1%, while methanol and other alcohols increase the signal because they reduce the surface tension and viscosity of the solution.

A twin-flame thermionic detector has been developed [212] by modifying a commercial device (Perkin–Elmer's 023-024). All the effluent from a microbore column is nebulized and fed into the primary diffusion flame. Then the combustion products are mixed with more fuel and fed into the secondary analytical flame, which burns in a glow regime. Since the two flames are spatially separate, the ion-current measurement only depends to a small degree on the nebulization, the solvent evaporation, and the combustion. At low volumetric flow rates, F = 1 μl/min, it also reduces the dependence of the detector's performance on the solvent type. In order to get the maximum mass detection sensitivity ($2 \cdot 10^{-11}$ g/sec), the gas deliveries to the two flames had to be optimized. McGuffin and Novotny [71] have described a flame ionization device that has a similar construction and also utilizes two flames.

All flame detectors produce a noisy signal, particularly when the solvent is organic. The signal must therefore be filtered with a relatively large time constant, and the slow response will make a large contribution to the widening of the peaks. On the other hand, the dead volume of the system may be reduced to a minimum by introducing the bottom end of the column directly into the flame. The large time constant of flame detectors makes them of less interest for fast capillary chromatography.

4.3.9. Radioactivity Detectors

Radioactivity tagging of biologically active compounds is an extremely effective method of studying biochemical processes involving them, such as their transport and metabolism. The more common tracers are the beta emitters ^{14}C, ^{32}P, and ^{3}H and the gamma emitter ^{125}I. The "homogeneous" detection schemes, which are now to a great extent out of date, involved adding a scintillation solution to the effluent after the column. The "heterogeneous" methods, which involve a cell filled with a solid scintillator, are now more popular. The usual cells for conventional chromatography are 60 μl in volume and are good for detecting beta emitters, although we do not know of any efforts to miniaturize this sort of cell for capillary chromatography.

A cell for detecting gamma emitters which lends itself to miniaturization is based on a crystal of sodium iodide [213], and its design is very simple. The microbore column's outlet is inserted into a hole drilled in the crystal. The crystal is placed in the plane of a photomultiplier, and the whole system is shielded with lead. The effective volume of the cell is determined by the dimensions of the capillary and can easily be reduced below a microliter. Ishii et al. [42-44] used a cell like this (volume 0.4 μl and measuring the activity with a Geiger counter) to detect a variety of radioactive nuclei. They obtained a detection limit of 2-20 ng. It should be noted that the chief reason for combining microbore columns and radioactivity sensors is not to gain high sensitivities, but to meet the safety requirements when working with radioactive compounds. The decisive aspects are the ability to get chromatograms without accumulating large quantities of radioactive effluent and the economic consideration that a column with only a few milligrams of sorbent may be simply destroyed after use.

Radioactivity detection is certainly not fully adapted to capillary chromatography. There is still the problem of detecting weak beta emitters such as tritium, which is the natural tracer for biochemical work, because tagging a biological molecule with ^{125}I may change its structure or affect its biological activity.

4.3.10. Postcolumn Derivatization

Reacting the compounds leaving a column to form colored, luminescent, or electrically active adducts that can then be sensed with an appropriate device is a widely accepted method for detecting enzymes (by activity), amino acids, sugars, metals, etc. Depending on the reaction and reaction conditions, the time needed for completion may range from several seconds to 20 min. The capillary reactor is a further source of extracolumn band spreading. There are three known types of flow reactor, i.e., open tubular capillaries, packed capillaries, and reactors with segmented flow. The problems involved in optimizing all three types of reactor have been well covered in the literature [214-218], including literature on capillary liquid chromatography [219, 220]. Therefore, we shall confine ourselves here to a simplified consideration of the problems of postcolumn derivatization. For a given reaction time the dispersion in the reactor can, in accordance with (3.51), be described by a variance

$$\sigma^2_{V,r} = \frac{1}{24}\left(\frac{t_r F_r r^2}{D_m}\right), \tag{4.25}$$

where $t_r = \pi r^2 \ell / F_r$ is the reaction time; r and ℓ are, respectively, the radius and length of the reactor capillary; F_r is the overall volumetric flow through the reactor ($F_r = F_c + F_{rg}$); F_c is the delivery rate of eluent into the column, and F_{rg} is the rate of reagent delivery.

Since the peak variance due to the spreading in the reactor is proportional to F_r^2, the addition of a flow reactor into a capillary chromatography analytical system is only admissible if F_c and F_{rg} are minimized. It must be kept in mind that in order to accelerate the reaction a significant excess of reagent must be added. Therefore, even for a maximum concentration of reagent in the feed solution diluting the effluent, too much in the reactor must be avoided. Usually, a two- to threefold dilution is permissible, i.e.,

$$F_{rg}/F_c = 0.5\text{-}1. \tag{4.26}$$

In this way the use of a flow reactor with a microbore column depends primarily on minimizing the eluent delivery F_c. It is also necessary to shorten to the limit the time taken to get the maximum yield of the derivative from the reaction. Arrhenius's law states that an increase in the temperature will increase the rate of the reaction; however, this method is limited by the thermal stability of the adduct, or for enzymes by their deactivation.

Let us consider the features of the postcolumn derivatization of amino acids with ninhydrin and using this for detecting amino acids in the effluent from a microbore column. This reaction takes 15 min at 100°C and 3.5 min at 130°C [221]. By optimizing the composition of the ninhydrin reagent [222] the reaction temperature can be raised to 160°C [223] and the reaction time shortened to 0.5 min. Under these circumstances a flow reactor (ℓ = 5 cm, r = 0.1 mm) can be used after a microbore column (d_c = 0.5 mm, L = 15 cm). The optimum elution rate is F_c = 2 µl/min (F_{rg}/F_c = 0.5), which ensures a negligibly small (less than 10%) contribution of the reactor to the overall variance. In order to improve mixing in the reactor it is recommended that it be filled with inert powder, such as nonporous glass beads 5-10 µm in diameter.

Postcolumn derivatization is possible for capillary chromatography because the permissible reactor volume is not so strongly dependent on the column diameter [see (3.54)].

The mixer design and the absence of pulses in the flows of effluent and reagent are essential to the efficiency of the mixing and the stability of the baseline. A good solution to the problem has been described by two groups [219, 223]. Postcolumn reactors have been successfully employed after microbore columns for reacting amines and amino acids with ortho-phthalic dialdehydes so they can be detected by fluorescence [172], and enzymatically inducing bile acids to form fluorophorescents [224, 225], and other compounds [220, 226]. Segmented-flow postcolumn reactors have not yet been applied to capillary liquid chromatography. However, the principle has been successfully used to design a postcolumn extractor [227] (ℓ = 100 cm, r = 0.15 mm). The cell contains a hydrophobic membrane separator for extracting the aqueous phase since it cannot pass through the membrane.

We shall now consider a very informative technique of postcolumn derivatization for analyzing bile acids on ultramicrobore columns. The fluorometric detection of a bile acid requires the oxidation of the 3-α-hydroxy grouping to the ketone group. This was done by the enzyme 3-α-hydroxysteroid dehydrogenase (3α-HSD). The β-nicotinamide adenine dinucleotide (NAD), which takes part in the reaction, was reduced to NADH, which could then be detected by a fluorometer, or electrochemically. The enzyme used in the reaction was immobilized on 200-400 mesh macroporous glass (CPG) using glutaric dialdehyde (a carrier with smaller particles would have been better). A PTFE or fused-silica reactor column (20 mm long, 0.34-0.5 mm ID) was filled with the enzyme bound to the carrier. The reactor column came after a separating column, which was made from fused silica (100-250 mm long, 0.26 mm ID) and filled with either Bilepack ODS silica gel or ODS SC-01 (both consist of 5-μm-diameter particles, JASCO). A 10 × 0.2 mm guard column was placed between the injector (V_S = 0.02 μl, JASCO) and the separating column and was filled with the same sorbent as the separating column. There are two ways of adding the NAD, viz., either before the separating column dissolved in the eluent or after it.

JASCO's FP-110 C fluorometer was used for the detector with our own flow cell, which had the dimensions 3.3 × 0.18 mm ID. The exciting radiation had a wavelength of 265 nm and the fluorescence was measured at 420 nm.

Since a fluorometer signal is proportional to the volume of the measuring cell, then miniaturizing for use with ultramicrobore columns substantially reduces the signal. However, if the signal is measured per unit sample mass, then the device Ishii et al. have created yields a ratio tenfold better than the one obtained for conventional liquid chromatography. This means that the threshold sample mass is a factor of ten smaller and the sensitivity of the analysis tenfold larger.

A comparison of the precolumn and postcolumn systems of adding the NAD showed that the baseline noise is lower for the precolumn system because of the attenuation of pulsations from the NAD feed pump. However, the precolumn addition of the NAD, and the salts needed to bring the pH into the optimal region for the enzyme, altered the k' for the bile acids. The postcolumn method has been developed to yield analysis of bile acids in biological liquids with a threshold sample mass of 0.04-7 ng.

In conclusion, we note that postcolumn derivatization for detection is unacceptable for fast microbore chromatography and doubtful for capillary technology (both in open tubular and packed columns) because of the difficulties in constructing flow reactors with volumes of a few nanoliters.

4.3.11. Other Sorts of Detectors

Many other relatively new sorts of detectors have been adapted for capillary liquid chromatography. We must mention the laser detector, which uses the thermal lens effect [228]; the atomic emission flame detector with inductively coupled plasma [229-231]; the catalytic pyrolyzer, which is specific for nitrosamines (the N—NO bond is split and the subsequent NO_2 is registered using chemiluminescence, a chemiluminescent detector with a thermal energy analyzer) [232]; the electrokinetic detector [233, 234]; the mass detector [235, 236]; laser polarimeters [237], which can be used for indirect measurements of light absorption at the 10^{-6} Å level; and, finally, chemiluminescence detectors with the postcolumn addition of peroxyoxalate [238]. The principles behind some of these detectors are covered in several reviews [239, 162, 163]. We shall now consider only the two most interesting.

The electrokinetic detector is based on an effect known from colloid chemistry. A potential and current are generated by charges being transported by a liquid due to the presence of a double electric layer. The current signal from this sort of detector is related [234] to the properties of the liquid and the interface between the two phases, i.e.,

$$I = qF = F\frac{\tilde{h}S_{sp}\tau_r}{1 + \tilde{h}S_{sp}\tau_r}2\Phi^*(C_L - C_F)\{1 - \exp \frac{La}{u}(1 + \tau_r)\}, \quad (4.27)$$

where S_{sp} is the surface area per unit volume of adsorbent, $\tilde{h}$ is the mass transfer coefficient, $\tau_r = \varepsilon_D/\bar{\kappa}$ is the relaxation time of the liquid, ε_D is the dielectric permittivity of the liquid, $\bar{\kappa}$ is the liquid's electrical conductivity, Φ^* is Faraday's constant, F and u are the volumetric and linear flow rates, L_a is the length of the adsorbent layer, C_L and C_F are the concentrations of the ions in the liquid and mobile phases, respectively, and q is the charge. The detector registers a current that is a function of the concentration of the ions in the effluent.

This sort of detector is a flow-sensitive detector in that the signal is proportional to the volumetric flow rate of the eluent. The principle can be used for on-column detection by measuring directly the electrokinetic current generated by the flow in the column capillary [234].

Unfortunately, the method is not universal, for it can only be applied when the mobile phase and the species being analyzed are partially dissociated, and when the electrical conductivity of the eluent is relatively low, i.e., $10^{-3}\ \Omega^{-1}\cdot m^{-1}$.

The mass detector, or evaporative analyzer, is especially interesting and is investigated in [240]. Many analytical problems, such as the determination of the molecular mass distribution of the peptides from a protein hydrolysate, have been difficult to solve because what was needed was a detector that could give equal responses to equal sample masses and not be affected by the chemistry of the various species. Although several detectors, such as the refractometer, IR spectrometer, and permittivity detectors, can sometimes respond equally to equal masses for homologous compounds, their range of application is limited. Thus, the discovery of a universal mass detector for high-performance liquid chromatography was eagerly awaited.

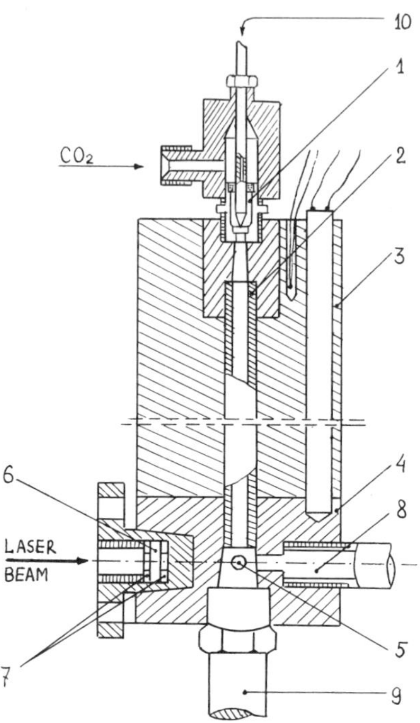

Fig. 4.16. Diagram of an evaporative analyzer for capillary liquid chromatography. 1) Nebulizer; 2) evaporative tube; 3) copper heater; 4) cell where scattered light is measured; 5) scattered light collector (glass rod) placed perpendicular to the laser beam; 6) glass window; 7) diaphragm; 8) light trap; 9) exit tube. The contribution of the detector to the band spreading variance $\sigma_{V,d}$ is 0.1 µl. The detection limit is 150 ng/sec. (Reproduced from [235] with the permission of Elsevier Science Publishers.)

The detector does not require any specially pure calibration standards. The way it works is illustrated in Fig. 4.16. The eluent is nebulized, and the colloid particles of the solute thus formed are passed in front of a light beam. The scattered light is detected by a photosensor placed at a certain angle to the beam. The linear range of the detector is relatively narrow, 10^{-4}-10^{-3} g/ml. The efficiency of solvent evaporation is increased by the delivery of hot gas to the evaporation tube.

An important virtue of this detector is that it can be used with gradient elution. Stolyhwo et al. [235, 236] have described a mass detector for capillary liquid chromatography. They used a helium—neon laser to illuminate the solute particles, and carbon dioxide to evaporate the solvent. The detector has been tested on triglycerides, n-alkylbenzenes, and the methyl esters of fatty acids. The detector's disadvantage is the impossibility of using buffering salts or other nonvolatile components of an effluent.

4.3.12. Liquid Chromatograph/Mass Spectrometer

The combination of a mass spectrometer and a liquid chromatograph is very promising. As is the case with gas chromatography, a mass spectrometer can relatively easily identify the compound being separated, and if the detection is set for a given m/z ratio, then it can select a desired component from a background of others in a chromatographic peak. Hence a mass spectrometer can be used as a universal detector by measuring the total ion current, or as a selective one by measuring a single ionic mass, while at the same time being a powerful instrument for determining the structure of the separated compound because of the information gleaned from the mass spectrum. However, in contrast to gas chromatography, there are a number of problems in connecting a mass spectrometer to a liquid chromatograph. These are related to the large volumes of gas or vapor that are produced when the liquid phase is evaporated, the presence in the eluent of nonvolatile or poorly volatile compounds (salts, buffer components, ion-pair reagents), the need to evaporate the solvent, and the difficulties involved in introducing poorly volatile or thermally labile compounds, which are typical of liquid chromatography, into the spectrometer.

The first task is to connect a liquid chromatograph and a mass spectrometer together and to create the requisite interface. This is an extremely difficult task because one of the devices works with a condensed phase (liquid) and the other requires a deep vacuum. A compromise solution is to provide the mass spectrometer with a powerful evacuation system (e.g., a cryopump [241]) and to reduce the volume of effluent either by splitting the stream (though this entails a loss of solute) or by using microbore columns which have flow rates of 10-60 μl/min. The latter alternative is obviously the most preferable because then the total sample is admitted into the spectrometer. It was also why the use of mass spectrometer as the detector arose early in the development of microbore chromatography.

It first appeared that all that was needed was to create a good interface to connect the microbore column to the ion source. In fact, the problem is much more complex, for in addition to the interface design and providing a powerful evacuation system the rapidity and sensitivity of the mass spectrometer itself have to be enhanced and the eluent has to be made more volatile without losing its selectivity. The latter problem is very difficult when it comes to biological objects because they are best separated by ion-pair

TABLE 4.6. Evaporation Properties of Several Liquid Chromatography Eluents (reproduced from [242] with the permission of Elsevier Science Publishers)

Type of chromatography	Solvent	ρ, g/cm^3	P^0, mbar	V_g, ml		ΔH, cal/g	T_b, °C
				at 25°C	at T_b		
Normal phase	n-Pentane	0.626	512	407	220	87	36
	n-Heptane	0.684	46	3630	208	87	98.4
	Benzene	0.879	95	2900	326	104	80.1
	Toluene	0.867	28	8200	298	99	111
	Methylene chloride	1.335	436	870	404	80.5	40
	Chloroform	1.492	197	1550	342	62	61.2
Reversed phase	Water	1.0	23.8	56000	1700	583	100
	Acetonitrile	0.786	84	5270	554	204	80
	Methanol	0.791	170.6	4096	685	263	65
	Ethanol	0.789	76	6617	493	200	78.5
	2-Propanol	0.785	58	6683	380	159.7	82
	THF	0.888	—	—	341	—	64
	Dioxane	1.034	53.3	6889	360	86.1	101

Note. V_g is the volume of the vapors at their boiling points (P = 1 atm) and at 25°C, for which $P = P^0$, and given the evaporation of 1 ml of solvent. $V_g(25°C) = V_g(T_b) \cdot 1020/P^0$ (P^0 in mbar); T_b is the boiling point, and ΔH the enthalpy of evaporation.

chromatography. The development of a liquid chromatograph/mass spectrometer thus requires the completion of several tasks:

1) the optimization of the spectrometer itself so that it can be used as a detector for high-performance liquid chromatography;

2) the optimization of the chromatograph to bring the eluent flow rate F in line with the capacity of the evacuation system and the time constant of the spectrometer;

3) the improvement of the eluents to increase their volatility or to use them as gas reagents for chemical ionization;

4) the design of an interface connecting the column and the spectrometer that can remove as much solvent as possible without distorting the form of the chromatographic peaks.

Like all destructive detectors, a mass spectrometer is a mass-sensitive detector and thus [242]: a) the signal falls exponentially to zero as the flow is slowed, b) the signal is not affected by the sample dilution, given that the product of the flow rate F and concentration C remains constant, c) the signal is increased by an increase in the effluent flow, and d) the chromatogram peak area is independent of F because, although the signal is proportional to CF, the peak width in units of time σ_t is inversely proportional to F. It is thus best to increase the material flux in the effluent by using the maximum sample mass and the fastest flow rate. However, doing the latter means that the capacity of the evacuation system must be raised or the molecular mass and density of the solvent increased, which unfortunately goes against our desire to increase the eluent volatility. Information on several of the eluents used for liquid chromatography is given in Table 4.6.

An alternative way to raise the capacity of the evacuation system is to develop an ion source that does not need a deep vacuum or one that can work at atmospheric pressure.

4.3.12a. Detection Limit of a Mass Spectrometer When Used as a Detector. A mass spectrometer does not have a very high sensitivity as a chromatography detector because of the low ionization efficiency and the loss of ions in the accelerating system of the spectrometer. As a result, a large number of molecules, $\tilde{N} \cong 2\cdot10^{10}$, have to be used to get a complete mass spectrum [242]. $\tilde{N}$ can be calculated from

$$\tilde{N} = \frac{k_m n'}{f_1 f_e f_i \tau_1} \approx 2\cdot10^{10} \text{ molecules/sec}, \tag{4.28}$$

where n' is the number of ions passing through the slit that are needed to determine their mass ($n' \cong 100$), f_1 is the fraction of ions that pass through the slit with scanning ($f_1 \approx 0.5$), f_e is the yield of ions from the ion source through the acceleration system ($f_e \approx 0.1$), f_i is the ionization efficiency of the ion source ($f_i \approx 10^{-3}$), τ_1 is the scanning time per unit mass ($\tau_1 \approx 10^{-3}$ sec), and k_m is a constant that takes into account the need to identify ions in small concentrations ($k_m \approx 10$).

The sensitivity can be increased to some extent, first by raising the ionization efficiency f_i. Take, for example, the ionization of haloaromatic compounds by electron impact ($f_i \approx 1$).

If we know $\tilde{N}$, then we can calculate the minimum mass of compound, $\tilde{m}_i$, that must be fed into a mass spectrometer per second to obtain a spectrum:

$$\tilde{m}_i = \frac{\tilde{N}}{N_A} M_i \approx 3.3\cdot10^{-14} M_i \text{ (g/sec)}, \tag{4.29}$$

where N_A is Avogadro's number. If M_i = 500 g/mole, then $\tilde{m}_i \approx 16$ pg/sec. This is the sensitivity if the spectrometer is used as a mass-sensitive detector and a whole spectrum is obtained. If selected ion monitoring (SIM) is used, the sensitivity can be increased by three orders of magnitude to around 1.5 fg/sec. We shall now determine the system parameters needed to get a sample mass-feed $\tilde{m}_i$ per second into a mass spectrometer.

If we take the concentration corresponding to the chromatographic peak's maximum C_M, then the flow of the i-th compound into the spectrometer will be

$$\frac{d\tilde{m}_i}{dt} = C_m F = \frac{\tilde{q}_i F}{(2\pi)^{1/2}\sigma_V} = \frac{\tilde{q}_i D_{m,i}}{N^{\frac{1}{2}} d_p^2 (1 + k')(2\pi)^{1/2}(h/\nu)}, \quad (4.30)$$

where $\tilde{q}_i$ is the mass of the i-th component in the sample. We can determine this from (4.30) as follows:

$$\tilde{q}_i = \frac{\tilde{m}_i N^{\frac{1}{2}} d_p^2 (1 + k')\sqrt{2\pi}}{D_{m,i}} \left(\frac{h}{\nu}\right). \quad (4.31)$$

Assuming $\tilde{m}_i$ = 16 pg/sec, $d_p = 5 \cdot 10^{-4}$ cm, $k' = 1$, $D_{m,i} = 5 \cdot 10^{-6}$ cm$^2 \times$ sec^{-1}, N = 10,000, h = 2.5, and $\nu = 5$, we get $\tilde{q}_i \approx 0.2$ ng.

This threshold mass sensitivity corresponds to the results of the latest work on liquid chromatography/mass spectrometry [121, 122, 243]. It is clear from the last equation that in order to reduce $\tilde{q}_i$ not only must $\tilde{m}_i$ be reduced, but d_p and k' should be decreased and D_m and ν must be increased. However, the latter is restricted by the time constant τ_{MS} of the spectrometer.

If the mass of the i-th component is a fraction β_i of the sample $\tilde{q}$, then the sample will be $\tilde{q} = \tilde{q}_i/\beta_i$. From (3.40) and (3.44) we know that $\tilde{q}$ is determined by the column parameters

$$\tilde{q} = \frac{\tilde{q}_i}{\beta_i} = 0.06 d_c^2 (1 + k') N^{\frac{1}{2}} h d_p \rho \varepsilon. \quad (4.32)$$

Substituting (4.31) into (4.32), we can solve for the column diameter best suited for the spectrometer:

$$d_c \geqq 6.6 \left(\frac{\tilde{m}_i d_p}{\nu \beta_i D_{m,i} \rho \varepsilon}\right)^{1/2}. \quad (4.33)$$

The reduced velocity ν, which controls $\tilde{q}_i$ and, according to (4.33), the diameter of the column, is determined by the time constant of the spectrometer.

4.3.12b. <u>Mass Spectrometer's Time Constant</u>. The time constant τ_{MS} is determined by the scanning time of the spectrometer. In a

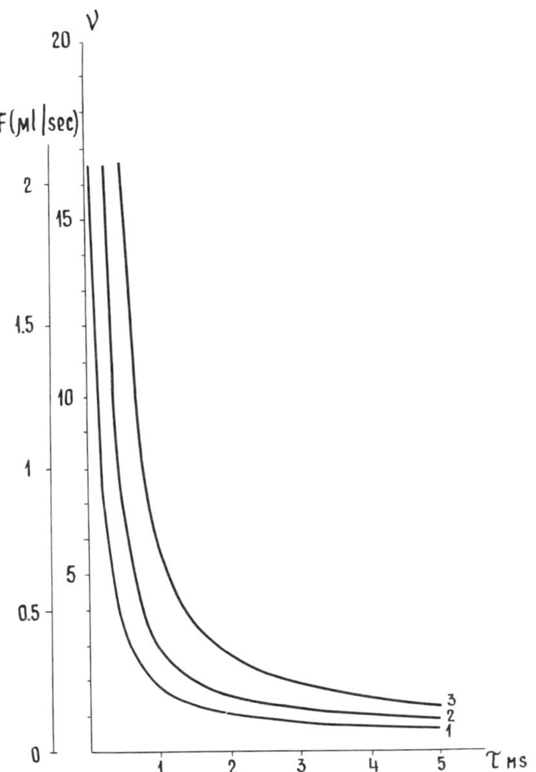

Fig. 4.17. Maximum reduced velocity ν versus mass-spectrometer time constant τ_{MS}. For various column efficiencies N: 1) 10^4; 2) $3 \cdot 10^4$; 3) 10^5. The flow rate F was calculated for d_c = 1 mm, ε = 0.8, d_p = 5 μm, $D_m = 10^{-9}$ m² × sec⁻¹, k' = 2, A = 1, B = 2, C = 0.05.

conventional magnetic spectrometer this time is more than 4 sec, but in the better magnetic and quadrupole spectrometers the value is 1-1.5 sec, which is the size of the time constant of a spectrometer when used as a detector for capillary liquid chromatography. We shall determine the maximum permissible value the time constant may be, $(\tau_{MS})_M$, as a function of ν and the other column parameters. If we use Kucera's estimate [14], then for the efficiency loss due to the detector's inertia not to exceed 5%, we must have

$$(\tau_{MS})_M = 0.3\sigma_t = 0.3\frac{t_R}{N^{\frac{1}{2}}} \tag{4.34}$$

and since

$$t_R = \frac{N d_p^2 (1 + k')}{D_m} \left(\frac{h}{\nu}\right) \tag{4.35}$$

and

$$\frac{h}{\nu} = \frac{B}{\nu^2} + \frac{A}{\nu^{2/3}} + C, \tag{4.36}$$

we get

$$\frac{3(\tau_{MS})_M D_m}{N^{\frac{1}{2}} d_p^2 (1 + k')} = \frac{B}{\nu^2} + \frac{A}{\nu^{2/3}} + C. \tag{4.37}$$

The last equation can be used to find the maximum reduced velocity as a function of τ_{MS}, N, d_p, and k'. Figure 4.17 gives $\nu(\tau_{MS})$ curves for various N, and the corresponding flow rates F.

Having found the reduced velocity from (4.37) and the column diameter from (4.33), we can find the elution rate F_{max} for the column that the spectrometer's evacuation system has to handle:

$$F_{max} = \frac{\pi d_c^2 \varepsilon u_{max}}{4} \approx \frac{0.5 d_c^2 \nu_{max} D_m}{d_p}. \tag{4.38}$$

Thus, Eqs. (4.29), (4.33), (4.37), and (4.38) yield all the parameters for a liquid chromatograph/mass spectrometer, i.e., $\tilde{m}_i$, d_c, ν_{max}, and $F_{max} = F$.

4.3.12c. Eluent for a Liquid Chromatograph/Mass Spectrometer. Clearly, a volatile eluent would present no difficulties. However, when buffer systems, ion-pair reagents, or "programmed" elution is involved, problems arise. The solution must lie in the use of isocratic or stepped elution [242], volatile acids and bases (Table 4.7) that decompose in the gas phase, or volatile ion-pair reagents. The salts of volatile bases and hydrochloric acid are also sufficiently volatile for use.

4.3.12d. Interface between a Liquid Chromatograph and a Mass Spectrometer. In order to connect the chromatograph column and the spectrometer together, the interface must provide the large sample yield acceptable in the spectrometer and be able to evaporate the least volatile samples. One other important requirement is that the interface produce little extra dispersion of the chromatographic zones (σ_V^2). The ratio of the flow in the column to that in the spectrometer is especially important to the design of the interface. Even though a chemical ionization (CI) source can handle 10-100 times more effluent than an ion source with electron impact (EI), the vacuum requirements of an ordinary mass spectrometer restrict the vapor flow rate to 20 ml/min. At the same time, conventional columns (d_c = 4 mm) operate with vapor flow rates of 2000 ml/min.

TABLE 4.7. pK Values of Some Volatile Acids and Bases (reproduced from [242] with the permission of Elsevier Science Publishers)

Compound	pKb	Compound	pKa
Ammonia	9.25	Acetic acid	4.75
Aniline	4.58	Benzoic acid	4.2
Diethylamine	11	Chloroacetic acid	2.85
Diisobutylamine	10.7	Cyanoacetic acid	2.45
Hydrazine	8.5	Dichloroacetic acid	1.48
Hydroxylamine	6	Formic acid	3.75
Pyridine	5.3	Phenol	9.9
Quinoline	4.8	Picric acid	0.40
Hydrocyanic acid	9.3	Thioacetic acid	3.33
Hydrogen sulfide	7.0		

In order to overcome the inability of ion sources to accept all the effluent, conventional columns must have a flow splitter and the sample must be concentrated. This complicates the construction of the interface and depresses the sensitivity of the spectrometer as a detector. A variety of interfaces have been described in the literature, and involve eluent transport using moving wires or bands [244, 245], direct introduction of the effluent via a capillary into a chemical ionization source with the partial evaporation of the solvent [246], membrane separators [247], or ionization at atmospheric pressure [248]. All these systems have limits with respect to the flow of eluent into the spectrometer, and hence can only be applied with conventional columns if the flow is split. This is the reason for the low yield of solute that reaches the spectrometer.

The thermospray technique [249, 258], which is a relatively new way of coupling mass spectrometers and liquid chromatographs, does not place such severe demands on the eluent flow rate and so can be successfully applied to conventional analytical columns (d_c = 4 mm). This may then be the only combination of mass spectrometry and liquid chromatography for which microbore columns are not much better than conventional ones.

The application of microbore, and particularly capillary, columns makes it possible to inject all the effluent directly from the column into the spectrometer, and to use the eluent vapor as a gas reagent. However, the reduction in the eluent flow rate limits the flow of compound into the spectrometer, which, as we mentioned above, has only a low sensitivity [242]. A very simple direct in-

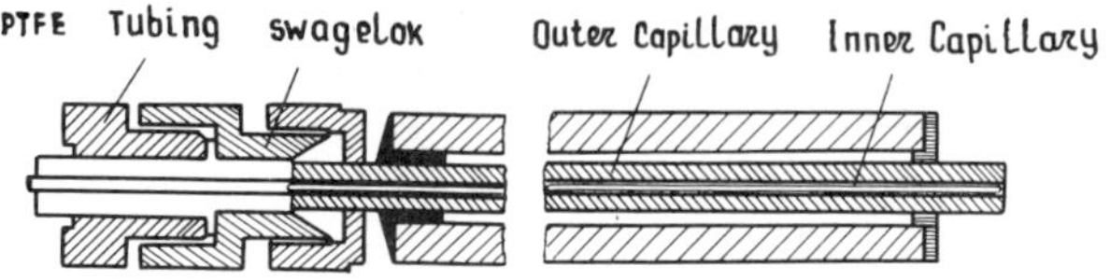

Fig. 4.18. Design of a capillary interface for the direct introduction of liquid. (Reproduced from [111] with the permission of Elsevier Science Publishers.)

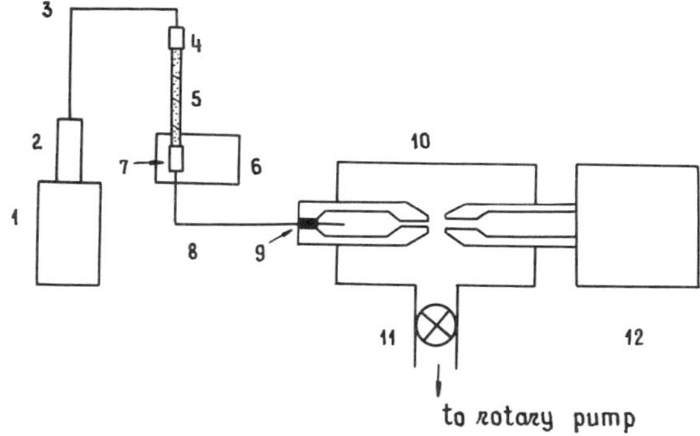

Fig. 4.19. Diagram of a jet separator to interface a microbore chromatograph and a mass spectrometer. 1) Syringe pump; 2) plunger; 3) PTFE tubing; 4) sample injector; 5) microbore column; 6) detector; 7) flow cell; 8) steel capillary; 9) silicone plug; 10) heated jet separator; 11) control valve; 12) mass spectrometer. (Reproduced from [109] with the permission of the American Chemical Society.)

terface is shown in Fig. 4.18 [111], and consists of a steel capillary with an internal diameter of 0.1 mm set into a second capillary (ID 0.25 mm) and into a tube with a 5-mm outside diameter. The interface is hermetically sealed to the lower end of the microbore column by a section of PTFE tubing and a Swagelok connector.

The main defect of this interface is that poorly volatile samples can clog up the capillaries, thus reducing the flow to the ion source and increasing the residence time in the interface. Moreover, even when a microbore column is used, the quantity of effluent is still too high for an electron impact source. Thus, there have been several attempts to introduce all the effluent, but first enriching the sample by applying jet separators [109] or vacuum nebulizers [110]. A jet separator designed for an effluent flow of 2-16 μl/min is given in Fig. 4.19. All the effluent is taken into a heated separator through a steel capillary and evaporated. The

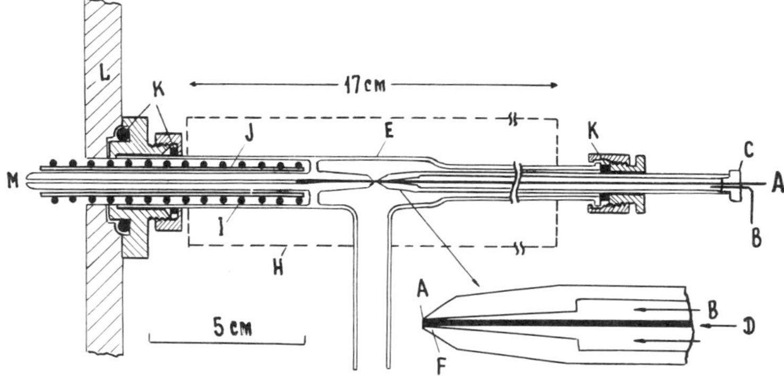

Fig. 4.20. Vacuum nebulizer interface for a mass spectrometer and a microbore column. A) Coaxial capillary tube; B) needle for injecting nebulizing gas (He); C) silicone plug; D) effluent; E) Pyrex nebulizing tube; F) to vacuum pump; H) furnace; I) heater of inlet capillary; J) heater screen; K) silicone O-ring; L) body of mass spectrometer; M) to MS's ion source. (Reproduced from [110] with the permission of the American Chemical Society.)

enriched sample is then passed into the chemical ionization source, where the solvent acts as a gas reagent. It is possible to maintain a constant pressure for different eluent flow rates by varying the speed of the evacuator of the separator.

Since the volume of the capillary is relatively small (4 μl), the interface cannot introduce any serious extra spreading of the zones. However, a jet separator is not really effective for poorly volatile compounds such as cholesterols, oligopeptides, or aromatic amines.

An interface employing a vacuum nebulizer is illustrated in Fig. 4.20 [110]. All the effluent from a microbore column is fed at 2-16 μl/min into the nebulizer through a steel capillary A (ID 0.15 mm) that has a wire with a mean diameter of 0.13 mm inside it. The coaxial exit of the capillary is placed in the center of the nebulizer chamber F, which is 0.4 mm in diameter. A nebulizing gas, helium, is fed in at atmospheric pressure and at a flow rate of 50 ml/min to form a vigorous stream in the narrow space between the chamber and the end of the capillary. The nebulizing head E is heated to 700°C. The distance between the nebulizer and the inlet orifice can be regulated to achieve the optimum vapor pressure in the chamber of the chemical ionizer. To avoid the adsorption of particles of poorly volatilized compounds on the input capillary, it, too, is heated to 700°C. An improved design of this interface is described in [112].

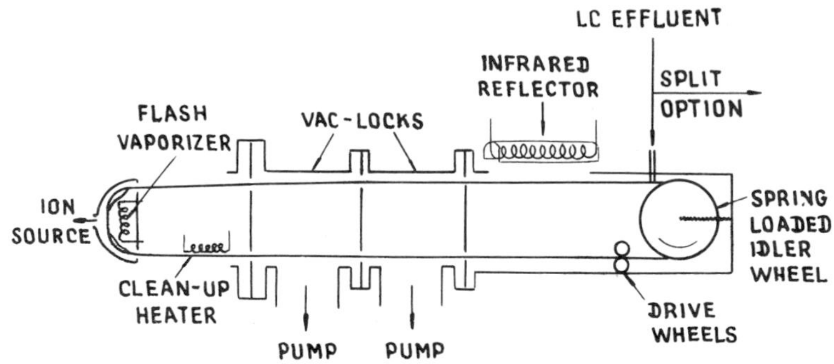

Fig. 4.21. Commercial moving-band interface. (With the kind permission of Finnigan Instruments.)

Unfortunately, neither jet separators nor vacuum nebulizers can enrich the sample sufficiently for electron impact sources.

Consequently, the method of directly introducing all the effluent from a microbore column into a mass spectrometer still requires a chemical ionization source. The interfaces described so far can accept most solvents, such as water, alkanes, alcohols, ketones, ethers, or nitriles, so long as their vapors are good reagents for chemical ionizers and produce few secondary peaks in the mass-number region of interest. Gradient elution may also be utilized, but the registration sensitivity will be dependent on the composition of the eluent.

It is also possible to connect liquid chromatographs and mass spectrometers by mechanically transporting the effluent on moving wires or bands [243-245], the mobile phase being evaporated during its passage through a system of vacuum seals. The sample can then be evaporated or pyrolyzed and fed into the chemical or electronic ionization source of a quadrupole mass spectrometer. A moving wire cannot transport more than 10 μl/min of effluent, i.e., all the effluent from an ultramicrobore liquid chromatograph column, a significant fraction of the effluent of a microbore column, and only 1% of the flow in a conventional liquid chromatograph.

A much more promising method is to transport the effluent on a moving band [243, 245] (Fig. 4.21). The capacity of this method is much higher, as much as 1 ml/min, and has a bottom limit of 0.05-0.15 ml/min for eluents containing a lot of water. Thus, a moving band can transport all the effluent of capillary, microcolumn, and conventional columns. Moreover, the transported samples are universally acceptable for every sort of mass spectrometer ionizer and every type of liquid chromatograph separator.

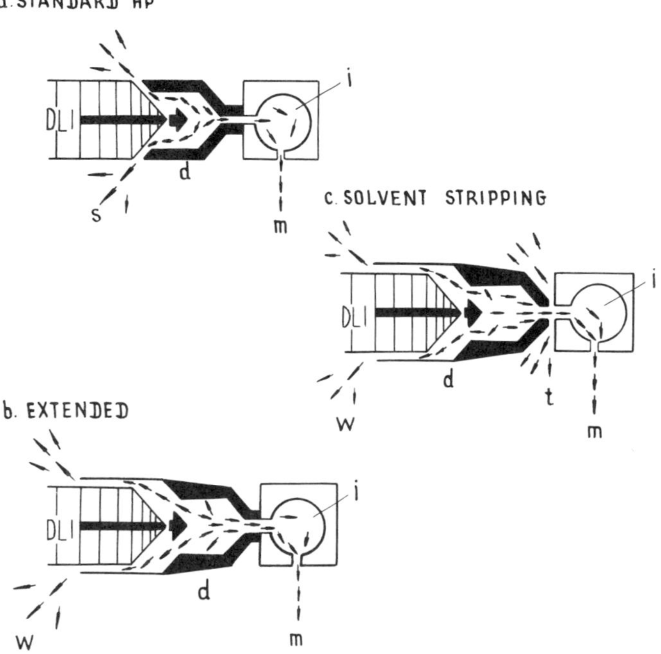

Fig. 4.22. Diagrams of direct injection interfaces with different types of desolvation chambers. HP) Hewlett–Packard Model 5985B option 04; d) DSC; i) ion source; m) ionized flow of solvent and solute molecules to the mass filter; s) flow of evacuated vapor between the probe tip and the DSC's seat; w) evacuated flow through the gap between the DSC's wall and the probe of the DLI; t) evacuated flow through the gap between the tip and the body of the ion source. (Reproduced from [119] with the permission of Elsevier Science Publishers.)

An interesting combination of a band interface and a fast atomic beam source (using fast Xe atoms) has been described in [250]. The paper showed how the method had been used to analyze peptides and oligosaccharides using liquid chromatography/mass spectrometry. Using a glycerin-containing eluent to create a glycerin matrix on the band, the detection sensitivity was tenfold higher than otherwise achieved.

It would seem that a moving band is the best method of connecting a liquid chromatograph to a mass spectrometer and meets all the specifications we set out above for such an interface. Moving bands have been made from steel or Du Pont's polyamide Kapton, which can be used to vaporize thermally labile samples at moderate temperatures. A diagram of a moving-band device is given in Fig. 4.21.

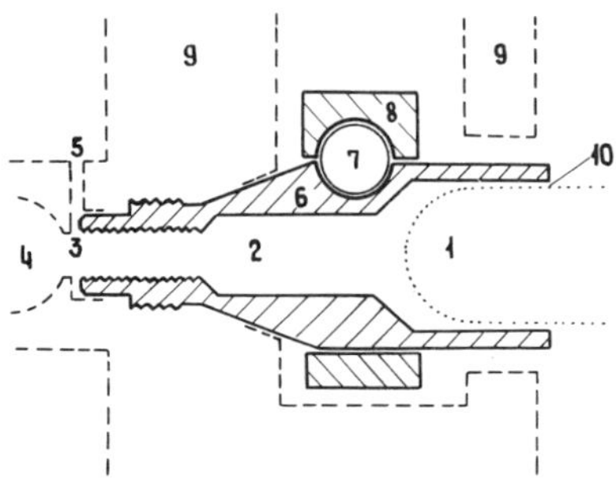

Fig. 4.23. Another solvent-stripping desolvation chamber. 1) Probe tip; 2) heated desolvation chamber and transfer line; 3) solvent-stripping annular pump gap; 4) ion source; 5) solvent-stripping pump line; 6) desolvating chamber wall; 7) cartridge with heater; 8) cartridge holder ring; 9) body of source and magnet; 10) controllable inner gap for splitting the flow. (Reproduced from [119] with the permission of Elsevier Science Publishers.)

This sort of interface can give rise to additional band spreading, and Yang [259] has shown that both the variance and the asymmetry factor depend on the speed of the band transporter, and may be changed by a factor of three.

Recently, workers have been trying to improve the performance of the direct type interfaces (DLI). Sugnaux et al. [119] have investigated whether a Hewlett—Packard 5985B gas chromatograph/mass spectrometer can be modified with an 04 device to give it liquid chromatographic capacity. They also compared the efficiencies of a standard desolvation chamber (Hewlett—Packard's HP-DSC) with modified versions, viz., a heated "extended DSC" and a heated "solvent-stripping DSC," the diagrams of which are given in Fig. 4.22. A better solvent stripper DSC is given in Fig. 4.23. This type of DSC can deliver a flow of 10-100 μl/min to a mass spectrometer if a powerful Cryopump evacuation system is used [241].

The problem of large eluent volumes being delivered to mass spectrometers when direct liquid introduction is being used is in matching the flow rate to the pressures needed in the ion sources and in creating stable diaphragms, which control the gas flow through the interface (for flows of 40 μl/min the orifice in the diaphragms must be around 4-6 μm). The pressure in the ion source can be regulated by changing the size of the slit between the input nozzle of the interface and the desolvation chamber. The tempera-

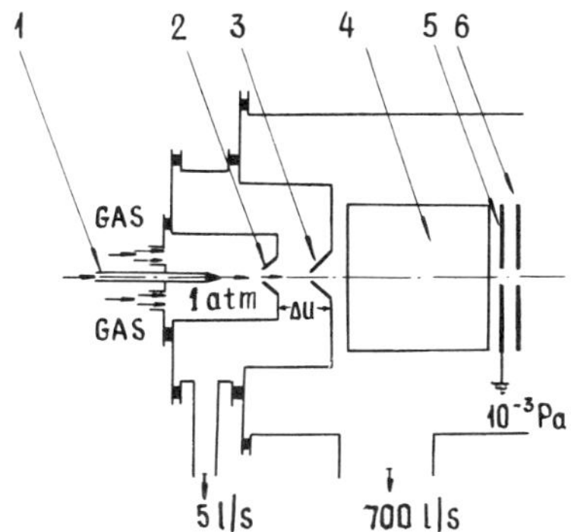

Fig. 4.24. Basic elements of ion source (ion extraction from a solution at atmospheric pressure). 1) Metal capillary for delivering the solution into the ionizer; 2, 3) system of diaphragms which form a gasdynamic jet. The capillary is placed coaxially with the orifices in the diaphragms, the first diaphragm acting as the antielectrode in the nebulizing system. A differential vacuum pump ensures the necessary pressure distribution in the various parts of the system. The parameters in the region between the diaphragms focus the ion stream onto the orifice in the second diaphragm. The potential difference between the diaphragms can be varied from 0 to 900 V. 4) Focusing lens; 5) inlet orifice to the mass spectrometer; 6) ion current monitor. (Reproduced from [252] with the permission of the authors.)

tures in the desolvation chamber and the source must also be controlled. At 150°C a pressure of 0.7 torr in the ion source will produce an effective negative chemical ionization, while 0.3 torr yields good positive ionization. At the same time, the vacuum in the desolvation chamber must be such as not to freeze the eluent in the interface capillary. Typically, a change in the temperature has a greater effect on a mass spectrometer than does a change in pressure. This is because the temperature affects the ratio of the ion intensities, while the pressure affects their overall intensity. The sensitivity of a mass spectrometer and the form of the peaks can be adjusted by regulating the ratio between the low- and high-molecular-weight clusters of the solvent. This interface provides stable conditions for soft ionization and subnanogram detection thresholds.

An interesting combination is that between a time-of-flight mass spectrometer, in which ionization is due to the nuclear fission of californium-152, and a liquid chromatograph [251]. A mass spectrum can be obtained from this device in about 1 min (m/z = 1-1000). The effluent fractions were transferred to the spectrometer by a rotating disk, and from 10^8 molecules per second were ionized. The system can be used to analyze large molecules, up to 10^5 daltons, even though the scanning rate is slow and ion output low. The system is very suitable for the chromatography of biological polymers.

An interface and ion source for a chromatograph/mass spectrometer that is of considerable interest is one based on a method developed in the USSR [252] called "the extraction of solute ions at atmospheric pressure." A solution of the solute in methanol, water, or acetonitrile, or a mixture of them, is fed at between 0.0001 and 1 μl/sec into a metal capillary. The capillary is kept at a potential of 2-4 kV relative to the diaphragm of the ion source (Fig. 4.24). The electric field causes the liquid flowing in the capillary to turn into a flow of charged droplets containing ions of the compounds in question. The multiple collisions the droplets have with gas molecules cause the solvent to evaporate from the droplets, which then lose their stability and disintegrate into smaller droplets as a result of the increased charge. The droplets finally end up as clusters, i.e., charged molecules surrounded by a solvent cloud. At a certain potential difference the cluster ions break up and a mass spectrum is obtained of the quasimolecular ions. The method is good for getting mass spectra of antibiotics, peptides, amino acids, nucleosides, steroids, and other complicated ionizable compounds, and in quantities of 10-100 ng.

Smith et al. [253] have shown that mass spectrometry is the ideal detection method for supercritical-fluid chromatography. The advantages are that more-volatile eluents can be used than are possible with conventional chromatography and that pressure programming and not gradient elution is used to regulate the process. If a chemical ionization source is used, the pressure in the source is kept stable by injecting extra gas. The columns used in this project were 10-30 m long and 100-200 μm in diameter, the stationary phase being either SE-52 or SE-54; and a quadrupole spectrometer with either an electronic or chemical ionization source was used. Isobutane, n-pentane, dichloroethane, or carbon dioxide was used as the mobile phase, and the investigation covered the separation of polycyclic aromatic hydrocarbons and oligostyrenes. The threshold mass sensitivity depends on the operation mode and is 100 pg for mass scanning and 1 pg for SIM.

4.3.12e. Chromatographic Columns and Separation Systems That Can Be Used with Microbore Liquid Chromatographs Coupled to Mass Spectrometers. Above we saw that conventional chromatograph columns can only be used with mass spectrometers by employing moving band interfaces or the thermospray technique. The most suitable column for direct liquid introduction is a microbore column with

d_c = 1 mm and F = 40-80 μl/min [119], using the eluents typical of reversed-phase chromatography. Applications with microbore columns and moving band interfaces have been described [243, 245, 259]. The smaller flow rates in microbore columns compared with conventional ones mean that the quantities of impurities precipitating in the mass spectrometer from the eluent are also smaller and, thus, the chemical noise level is lower.

Some separate work has been done on connecting mass spectrometers to open tubular [69] and ultramicrobore packed columns [122]. Since the flow rate F is <1 μl/min, all the effluent can be directly injected into an electron impact source. Understandably, no sensitivity payoff in terms of lower threshold sample mass has been observed. Incidentally, Niessen and Poppe [69] were able, by using a makeup flow to form a jet at the outlet of the diaphragm, to achieve a mass detectivity threshold of 1-10 pg for open tubular columns. The standard deviation for the extracolumn spreading $\sigma_{V,ec}$ was no more than 1 nl.

Clearly, a fast chromatograph (F = 3-6 ml/min) cannot be connected to a mass spectrometer both because of the impossibility of injecting into the spectrometer the requisite volume of effluent, and because the spectrometer's time constant is too large to be consistent with the speed of the chromatographic process.

The substances analyzed and separated in the literature on liquid chromatograph/mass spectrometers include aromatic hydrocarbons and amines, steroids, amino acids, oligomers, and medicinal preparations, including some that were contained in biological fluids. These are all described in reviews [242, 254-256]. The experiments covered were generally carried out to test an interface or chromatograph/mass spectrometer system that had been developed, and not as part of an attempt to use the system in routine analysis. Thus, we must say that these systems are still research objects and not standard systems available for routine use. Nevertheless, several firms are marketing interfaces for their spectrometers; e.g., Finnigan Mat (USA) and V. G. Analytical (UK) offer moving band systems, and Hewlett—Packard (USA) offers a direct liquid introduction device. There are now some 100 such systems described in the literature [257].

In order to obtain useful analytical systems that can operate in the optimum regimes of the chromatographs and rapidly analyze samples which may contain trace components, mass spectrometers must be developed such that

1) their time constants are reduced to 0.2 sec;

2) their sensitivities are enhanced to 10^6-10^7 molecules/sec (by increasing the ion output of the source and speeding up the scanning rate);

3) the devices can be independently tuned either to high sensitivity or to get short time constants;

4) the buffer solutions and ion-pair reagents are volatile or decompose when heated;

5) the interfaces, ion sources, and evacuation systems will produce ionization that is either soft (to yield molecular ions) or hard with minimal overall spreading at flow rates up to 100 μl/min.

4.4. DEVICES FOR INJECTING AND CONCENTRATING SAMPLES

In order that the efficiency losses due to the sample injection remain below 5%, the sample volume should satisfy Eq. (4.5). It can be seen from Tables 4.1 and 4.2 that for packed microbore columns with a d_c of 1 mm the limiting permissible sample volumes are from 0.3 to 1.4 μl, depending on the column length and eluent flow rate. Note that when ultramicrobore packed columns are being used (d_c = 0.2 mm), the demands on the sample size are much more severe, i.e., from 10 to 60 nl. If the column diameter is further diminished, then the size of the sample becomes very difficult to reproduce using modern techniques. Thus, the sample volumes for open tubular capillary columns (d_c = 30 μm) must be 13 nl or less (see Table 4.3). However, Knox and Gilbert [22, 23] and Yang [259, 260] have shown that open tubular columns must be 10 μm or less in diameter in order for them to exceed packed columns in terms of separating ability and analysis speed for high-performance chromatography (d_p = 5-10 μm). Table 4.3 demonstrates that the permissible sample volume for an open tubular column with d_c = 10 μm is only 0.5 nl.

For the present, the problem of accurately and reproducibly injecting a sample for packed microbore columns with diameters of 0.2-1.0 mm has been solved by using rotary valve injectors, which have loop capacities of 0.02-1 μl [261, 262]. Although a variety of ingenious methods for introducing samples with volumes less than 20 nl have been suggested [17, 20, 108], injecting such volumes requires an extremely skillful experimental approach, and a great deal of standardization remains.

Fortunately, it turns out that in many cases significantly larger volumes than those given in Tables 4.1-4.3 can be injected into the columns. In fact, as we showed in Section 3.3.2 when sorption is the separation mechanism, as it is for most chromatographic techniques other than steric-exclusion chromatography, the zone constriction effect, which occurs in the top of the column when the sample is introduced (dissolved in a solvent that ensures a high sorptive capacity), ameliorates the situation. This is the basis for the sample concentration schemes that we shall now describe.

Historically, the first and simplest injection procedure for microbore columns was developed by Sandakhchiev, Kuz'min, and Grachev [7-10], who were followed by Ishii [37, 38], the sample being injected when the flow was stopped and the column disconnected. The method has been proved on packed microbore columns with diameters of 0.5 mm [38, 39], packed ultramicrobore columns with diameters of 0.1-0.15 mm [45], and on open tubular columns with diameters of 50-60 μm [74, 78], samples as small as 20 nl having been achieved. The requisite sample is sucked into a steel capillary (0.1-0.3 mm ID) or into a fused-silica capillary using a graduated syringe micropump. Then a small quantity of the mobile phase is added to the capillary, which is then inserted into a PTFE column or connected via a short piece of PTFE tubing to a capillary column. Immediately after the connection has been made, the syringe pump with the eluent is turned on to start the discharge regime.

This method, too, has drawbacks. For example, nowhere in the reports written by the Nagoya University Group headed by Ishii does it mention what pressure the upper section of the PTFE column can stand with this type of sealing. Evidently, pressures above 10 MPa are not possible. Furthermore, regularly resealing a column increases the probability of leaks and depresses the quality of the packing. Nor is the method convenient for the operator, and the moment at which the elution starts cannot be fixed accurately. Hence, loop rotary valve injectors are preferred for quantitative analyses and reliable long-term work. Loop injectors for packed microbore columns are now made by many manufacturers, such as Rheodyne, Inc., Valco Instr. Co., JASCO, and EM Science; and Rheodyne's 7410, 7413, and 7520 injectors are very consistent sampling devices ($\sigma_{rel} \leqq 0.05\%$) for columns with diameters of 1 mm. The sample chamber of the 7520 is an accurately reamed out orifice in the rotor of the valve (an internal loop). The valve comes with several replaceable rotors for different volumes, viz., 0.2, 0.5, and 1.0 μl. The volumetric variance at the end of the outlet capillary (ID 0.13 mm) is less than 0.2 μl^2. The loop is guaranteed to be completely filled by a 10-μl syringe, and its accuracy is also guaranteed.

The rotary valve injectors (internal loop) with the smallest sample volumes are those made by Valco (70 nl) and JASCO (the ML-422, 20 nl). The latter device is described by Takeuchi and Ishii [262], and in spite of its very small sample volume, the volume reproducibility is very good, with a peak-height relative standard deviation σ_{rel} of less than 0.6%.

When making a rotor-type valve injector, the channel must be accurately reamed out and the rotor fixed in position; otherwise, stagnant zones may develop in the device, leading to additional spreading in the peaks and their asymmetry. The dead volume between the sampling valve and the column must be reduced to a mini-

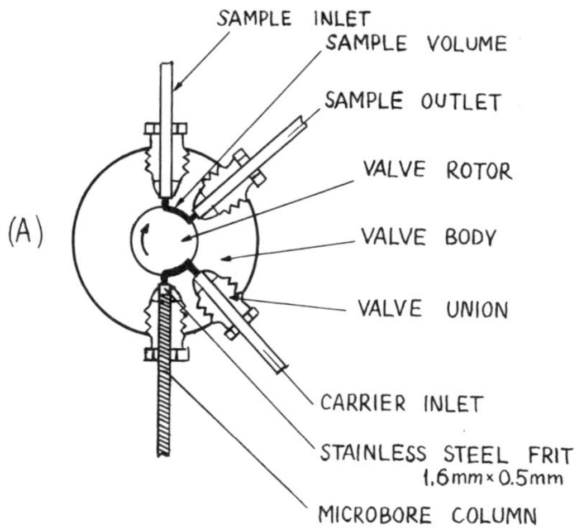

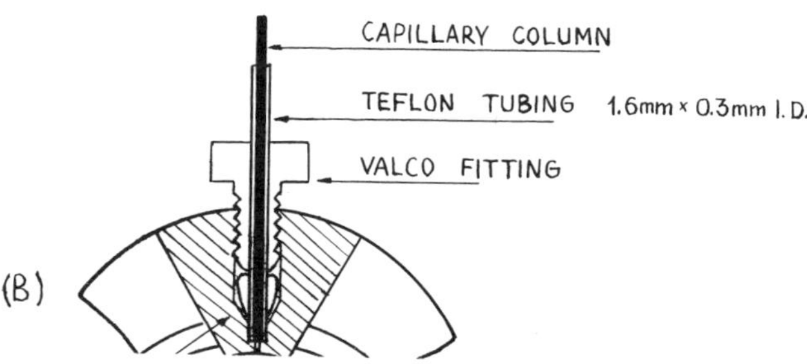

Fig. 4.25. Direct connection of Valco rotary valve injector (internal sampling loop) with A) a stainless steel microbore column [14] and B) an open tubular column [263]. (Reproduced with the permission of Elsevier Science Publishers.)

mum. Stainless steel columns (OD 1/16") can be screwed directly into the injector (Fig. 4.25A), using appropriate couplings with ferrules for sealing. The techniques for joining loop injectors to ultramicrobore, fused-silica, or open tubular columns are described in several reports [261, 263], and shown in Fig. 4.25B.

In order to introduce very small samples ($V_S \leqq 20$ nl), which is necessary for open tubular columns, packed capillary columns with irregular packing, and ultramicrobore packed columns, a variety of intricate methods have been invented. Split sample injection, the standard method for gas chromatography in capillary columns, should be mentioned first. The eluent stream containing the sample is split into two at the column inlet. Most of the mass of

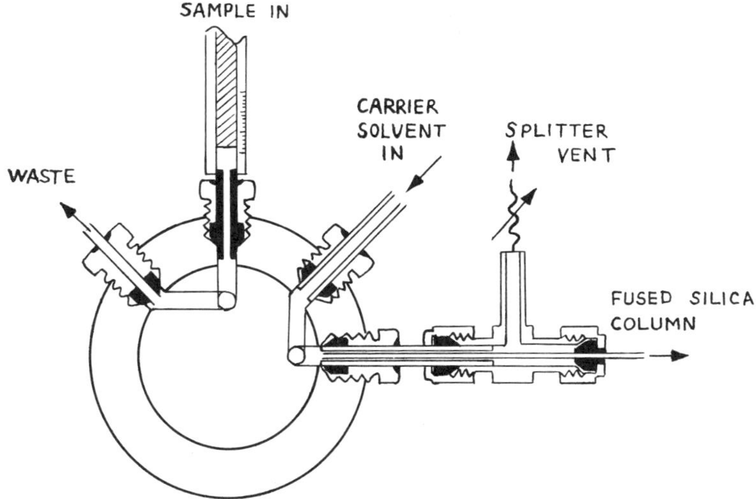

Fig. 4.26. Split flow injector for capillary columns. A modification of a Valco rotary valve with an internal sampling loop, capacity 0.5 µl. (Reproduced from [17] with the permission of Elsevier Science Publishers.)

liquid is removed as waste through the device that regulates the hydraulic resistance (a valve or restricting capillary). Thus, only a small fraction of the injected samples is allowed into the column. The size of the sample fed into the column is calculated from the size of the injected sample and the ratio between the flow rates in the column F_C and in the waste channel F_W.

Tsuda et al. [24, 87] have investigated syringe injection in combination with a splitter for pressures up to 10 MPa. At higher pressures the sample becomes more difficult to inject using a microsyringe, and acceptably accurate volumes could only be obtained for very large splitting ratios, i.e., $F_W/F_C > 10^4$. Yang [17] has suggested a more successful split flow injector. It is shown in Fig. 4.26 and is based on a rotary valve with an internal measuring loop. The system yields reproducible peak areas (with a relative standard deviation less than 2.1%) and can be used with pressures up to 50 MPa.

McGuffin and Novotny [108] have developed the split flow technique, as shown in Fig. 4.27, and proved their device on open tubular columns, getting samples down to 1 nl. In addition to the flow-splitting technique, McGuffin and Novotny also employed what they called a "heart-cutting" method to reduce the sample volume. The purge valve is initially open so that no eluent reaches the column. After the sample has been injected via a conventional loop rotary

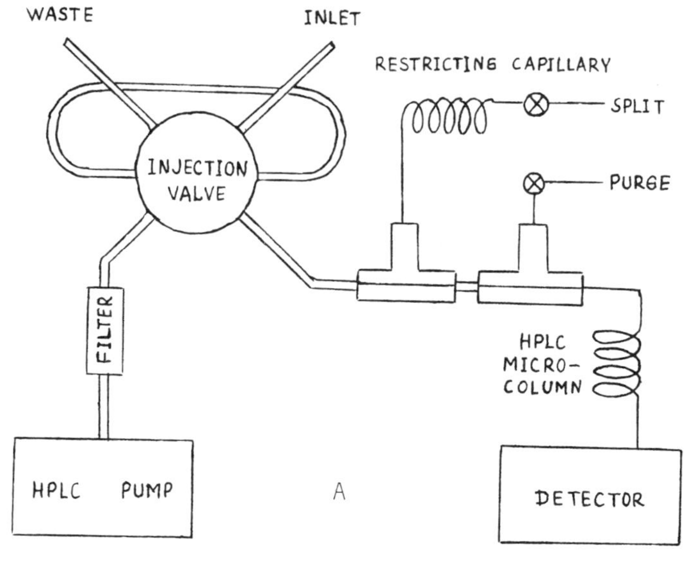

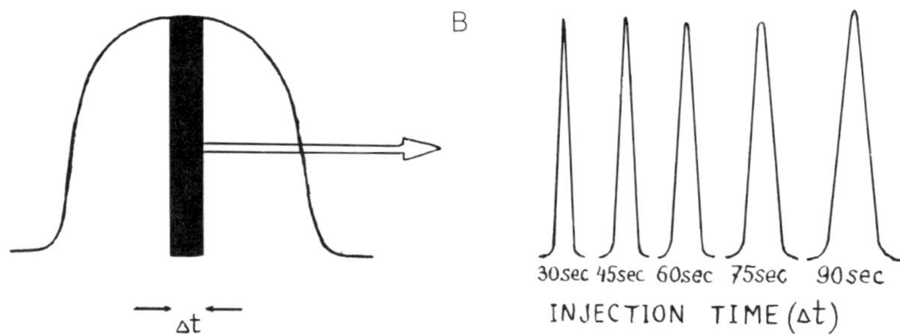

Fig. 4.27. A) Split flow injector with "heart cutting" for capillary columns; B) form of peaks as a function of sampling time. (Reproduced from [108] with the permission of the American Chemical Society.)

valve, the purge valve is shut for a certain period of time Δt, during which the center of the sample plug enters the capillary column. Thus, the volume of sample entering the column is $V_S = F(B_C/B_W)\Delta t$, where F is the overall volumetric flow rate, and B_C and B_W are the hydraulic permeabilities of the column and the waste channel, respectively. After the time interval Δt has elapsed, the purge valve is reopened so that the remainder of the sample can be removed. In this way the initial sample volume is significantly greater than the cut-out fraction. The instants at which the purge valve is opened and shut can be ascertained reasonably accurately, and in fact are determined empirically and depend on F, B_C/B_W, and the di-

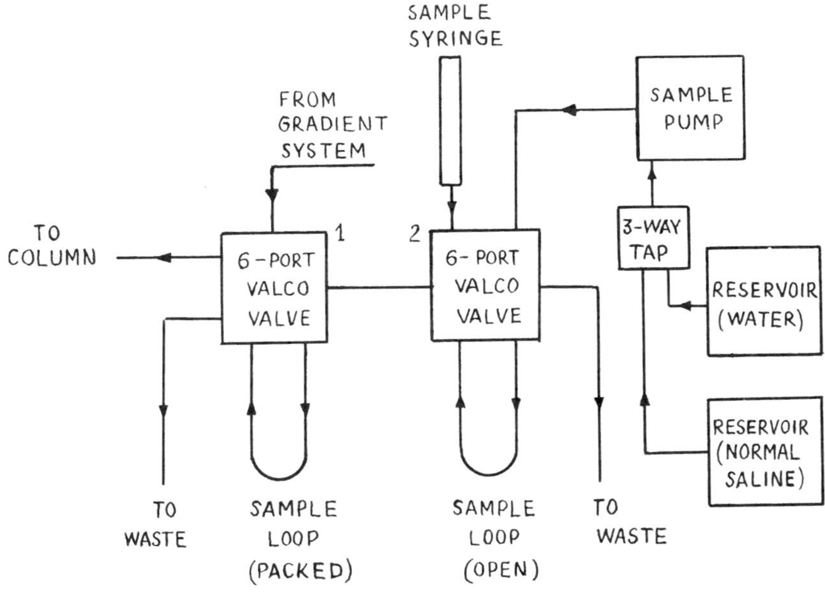

Fig. 4.28. Block diagram of sample enrichment system using two Valco loop valves. (Reproduced from [53] with the permission of Elsevier Science Publishers.)

mensions of the connecting capillaries. Even when a sample of 2.4 nl is injected manually over a period of 10 sec, the absolute error (ΔV_S) is less than 0.15 nl.

In principle, this system could be automated, and it is the best, to date, for introducing very small samples (less than 20 nl). Unfortunately, it has all the disadvantages of split flow injectors, viz., the loss of most of the sample mass as waste, and instabilities over time of the splitting ratio due to gradual changes in the permeabilities of the columns and the restrictor. The latter drawback can be partially overcome by introducing an internal standard with the sample.

Another simple approach is worth mentioning and allows subnanoliter samples to be manually introduced into open tubular columns [20, 83]. A short (few millimeters) section of the top of an open tubular column is quickly heated with a small flame and immediately immersed in a solution of the sample. The solution is sucked up by capillary action into the space vacated by the evaporation of the mobile phase from the column. Given that a 5-mm length of the capillary is vacated and that the column is 10 μm in diameter, the sample volume would be 0.4 nl. This procedure requires great experimental mastery, as do all of the operations involved with open tubular capillary columns.

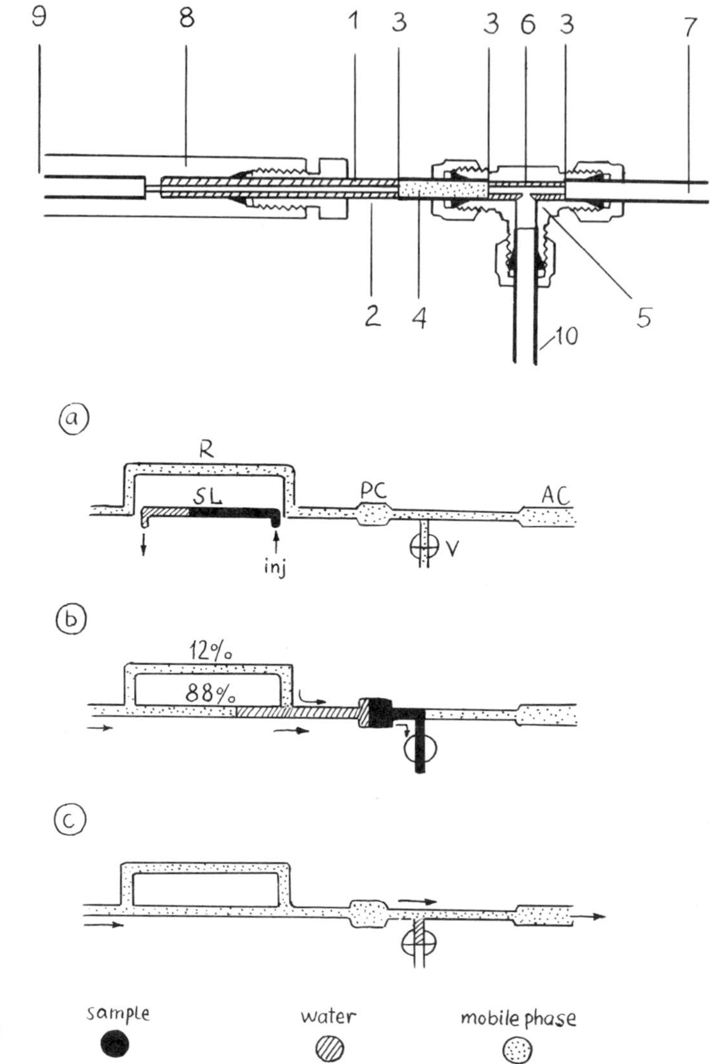

Fig. 4.29. A) Design of sample enrichment precolumn: 1) precolumn, 33 mm × 1.1 mm; 2) PTFE capillary; 3) filter; 4) sorbent in the precolumn; 5) Swagelok T-junction; 6) PTFE capillary with V-shaped groove; 7) analytical column; 8) Swagelok 1/16"-1/16" connector; 9) to sampling valve; 10) to valve V. B) Diagram of sample concentration in the precolumn: SL is the measured volume (of the loop), R is the restrictor, PC is the precolumn, AC is the analytical column, and V is the valve. (Reproduced from [165] with the permission of Elsevier Science Publishers.)

Whatever injection technique is employed, it is recommended that for microbore liquid chromatography the sample be dissolved in a solvent that promotes the sorption of all the sample components, so that the sample is concentrated in the upper section of the column. One such solvent for the alkylsilane hydrophobic sorbents is water. In order to increase the solubility of nonpolar solutes, 10-30% of either alcohol or acetonitrile can be added to the water. Sample concentration is not only useful for reducing the spreading due to the injection technique, it is essential for trace analyses, which is the most common task when dealing with biological fluids, tissue extracts, and environmental samples.

In order to enrich and accumulate a sample, it is often best to use a precolumn that is filled with the same sorbent as the main column, or a sorbent that has the same functionality but lower sorptive capacity. The sorbent grains in the precolumn may be significantly larger in diameter (as much as 100 μm) than those in the analytical column. This both reduces the hydraulic resistance of the column and facilitates the removal of insoluble particles from the sample when the precolumn is washed.

The most successful systems in the literature [40, 53, 165, 196, 264] on enriching samples are those shown in Figs. 4.28 [53] and 4.29 [165]. In the first of these two systems the precolumn, made in the form of a loop of a six-port Valco rotary valve, is first delivered the requisite quantity of sample by an auxiliary pump. The precolumn is then washed with water via a three-way tap in the pump inlet line and another Valco valve in the discharge line. The precolumn now contains the sample as a narrow zone of adsorbate in contact with water. Finally, the first valve is turned to connect the precolumn with a gradient device and the analytical column.

The device shown in Fig. 4.29 is significantly simpler and needs neither an auxiliary pump nor a tap. A Rheodyne Model 089-0932 valve is used to deliver the sample, the valve having a 175-μl loop and a divider which sends 12% of the mobile phase through a channel (R) that avoids the loop. Some (100 μl) of the sample, dissolved in water, is introduced into the loop, which is first filled with water, and after valve V is opened 200 μl of the mobile phase is allowed to flow through it. Since only 12% of the mobile phase flows through channel R, the eluent (50% methanol) is diluted in the precolumn to 6% in methanol, and this ensures the sample enrichment effect takes place. The 75 μl of water left in the loop is used to wash the precolumn. After shutting off valve V, the precolumn is connected to the main column. In order to eliminate dead zones in the Swagelok T-junctions, a PTFE capillary (ID 0.3 mm) with a V-shaped groove is inserted in the middle.

Note that although Scott and Kucera's enrichment system (Fig. 4.28) is less simple than the system in Fig. 4.29, it has an indisputable advantage in that the extent of the sample concentration can be controlled.

If the sample to be enriched does not have to be cleaned of mechanical and water-soluble impurities, then it can be concentrated in the analytical column itself. This was demonstrated for a DNS amino-acid system by Maltsev et al. [12].

4.5. SOLVENT DELIVERY SYSTEMS

The usual eluent flow rate through 1-mm-diameter packed microbore columns filled with a stationary phase whose particle diameter is 5 μm must, according to Tables 4.1 and 4.2, lie between 20 μl/min (superefficient analysis) and 350 μl/min (high-speed analysis). The corresponding rates for ultramicrobore packed columns (d_c = 0.2 mm) are 1 μl/min and 14 μl/min, while Table 4.3 shows that for open tubular columns (d_c = 10 μm) the eluent flow rate may need to be as slow as 2.5 nl/min. The lowest solvent delivery rates (100 μl/min) typical of most commercial liquid pumps have been reduced in recent times to 10 μl/min, mainly to meet the needs caused by the development of microbore liquid chromatography. Solvent delivery systems that can be used to realize both isocratic and gradient elution regimes are now well developed and have been incorporated into commercial devices for microbore columns.

The two approaches taken in the design of equipment for capillary liquid chromatography, which we mentioned in Section 4.2, are especially apparent vis-à-vis solvent delivery. The tendency for firms to unify their solvent delivery systems has dominated with most firms preferring to modify their existing equipment, instead of developing new models specially for microbore columns. Moreover, they have been successful in this, and reciprocating pumps have been created with stable liquid deliveries of 10-100 μl/min. These pumps are excellent for isocratic regimes and adequately accurate for gradient elution with 10-90% of the final solvent [52, 53, 67, 158]; moreover, all the systems retain the standard pump parameters for the solvent rates required by conventional high-performance chromatography, i.e., 0.1-10 ml/min.

One such universal system includes the Waters 590 programmable module (flow rate range 1 μl/min-45 ml/min) for isocratic elution or two Waters 510 pumps in combination with the Waters 680 gradient controller for a gradient regime [158]. The latter system produces a gradient regime between 45% and 80% of the final solvent in which the residence time and peak areas are outstandingly reproducible, the relative standard deviations being lower than 1.3% and 1.8%, respectively. The accuracy and stability of the solvent delivery of the dual piston pump are obtained here by a system of noncircular gears and a microprocessor-controlled stepper motor which divides each microliter of solvent into 48 discrete steps.

Scott and Kucera [52, 53] have developed a way of modifying reciprocating pumps to reduce their deliveries to a 1-10 μl/min range. They replaced the frequency generator of a Waters 6000A pump by a Hewlett–Packard 3311A function generator, and so were able to reduce the minimum switching frequency of the pump to 2 MHz and use two pumps per gradient by modifying a Waters M 660 programmer or a Schoeffel Klik 1 control system [53].

Gilson Medical Electronics offers its M 302 pump, which has an accurately controlled stepper motor (1200 steps per revolution), with interchangeable liquid heads. This enables the pump to be used for deliveries from 0.005 to 100 ml/min.

Although modern reciprocating pumps have been successfully applied to packed microbore columns with flow rates of 10 μl/min, this seems to be their limit because the check valve of a reciprocating pump may have a normal leakage of 0.1-1 μl/min, and it is difficult to reduce the volume delivered by a plunger to below 30 μl per stroke. Thus, it is a problem to use reciprocating pumps for microbore columns with diameters of 0.5 mm, and they cannot be considered for smaller diameters without ancillary devices. Moreover, we shall show later that reciprocating pumps are not ideal when using gradient elution even for column diameters of 1 mm. The more promising approach, therefore, seems to be the development of special-purpose syringe pumps for capillary chromatography and gradient systems, even if only a few firms are progressing in this direction, e.g., Brownlee Labs, ISCO, and the Science and Technology Organization of the USSR Academy of Sciences. Syringe pumps were used in the pioneering studies of microbore chromatography [7-10], but they seem to be flexible enough to provide all the requisite flow rates for packed microbore and open tubular columns. The syringe must be large enough to contain sufficient eluent for at least one analysis. A 0.5-2 ml syringe is adequate for a microbore column 0.5-1 mm in diameter [12, 13, 38], while only 50-200 μl are needed for open tubular columns.

The inaccuracy in the solvent delivery of a syringe rises with the diameter of the plunger. It is important, therefore, both that the plunger movement be made as accurate as possible, and that the plunger be long so that its diameter can be reduced as much as possible. The delivery rates of syringe pumps are fairly independent of the pressure and solvent viscosity. However, a constant rate is not achieved instantaneously, the time needed to do so being dependent on the compressibility of the liquid. Although the pumps have to be refilled with eluent after each analytical cycle, the ease with which the refill can be done and the absence of rate fluctuations are important advantages.

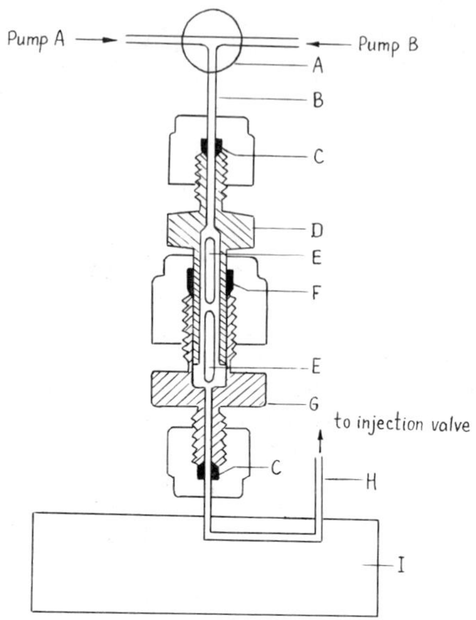

Fig. 4.30. Diagram of a dynamic micromixer. A) A Valco 0.01" valve; T) the mixer; B, H) steel capillaries (ID 0.007"); C) a Swagelok 1/16" cone seal; D) a Swagelok 1/8" × 1/16" reamed-out adaptor; E) Teflon-coated moving rods; F) 1/8" Swagelok seal; G) a Swagelok 1/8" to 1/16" adaptor with a zero-volume dead zone; I) a magnetic agitator. (Reproduced from [67] with the permission of the American Chemical Society.)

The plungers may be driven with either a synchronous or a stepper motor, and because the plunger diameter is small, so, too, is the load on it. For instance, the load on a plunger with a cross-sectional area of 4 mm^2, syringe volume of 250 μl, and at a pressure of 100 kg/cm^2 is only 4 kg. Since a syringe pump can be operated both to push liquid out and to suck it in, it can be used for sample injection by the stop-flow technique [37, 38].

The syringe pump systems now marketed by several firms for capillary chromatography realize the potential of the technique. For example, the μLC-500 system made by ISCO can generate pressures of 0.1-70 MPa at flow rates of 0.02-600 μl/min and with a stability of better than 1% and accuracy of 1%, the syringe holding 50 ml, which is enough for 20-200 analyses.

There is yet another, more subtle advantage of syringes: they work equally well at high and low pressures. The operational efficiencies of the ball-type check valves in reciprocating pumps are very dependent on pressure.

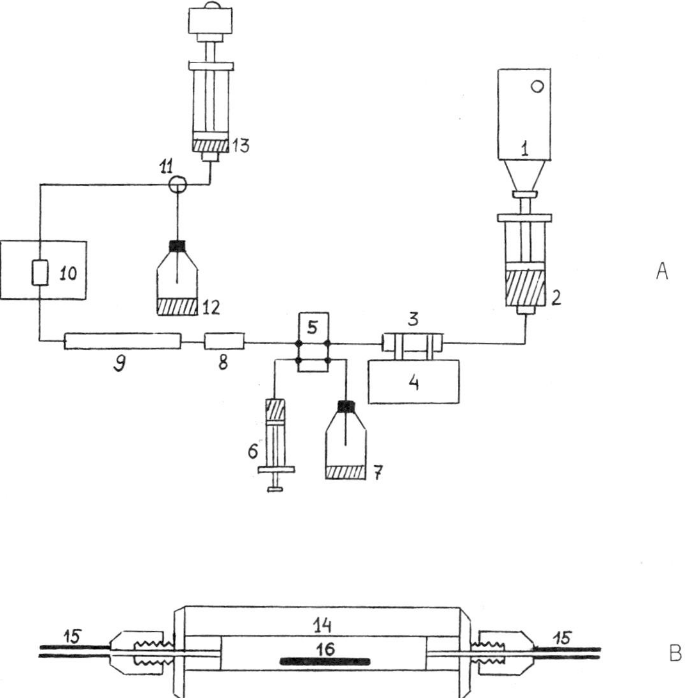

Fig. 4.31. A) Continuous gradient elution apparatus: 1) micropump; 2) syringe pump; 3) mixing chamber; 4) vibrator; 5) injector; 6) sample; 7, 12) overflow reservoir; 8) precolumn; 9) analytical column; 10) detector; 11) three-port valve; 13) pump for creating the requisite pressure. B) Design of mixing chamber: 14) gastight syringe; 15) needle; 16) 10 mm × 0.63 mm (OD) steel rod. (Reproduced from [107] with the permission of Elsevier Science Publishers.)

Brownlee Lab's MPLC Micropump [265] is exemplary for capillary chromatography. The system uses two microprocessor-controlled syringe pumps with automatic refill valves, the refilling time being about 30 sec. The accuracy of the delivery in the gradient mode is achieved by dividing each microliter into 2400 substeps that can be proportioned between the two pumps. A short column filled with glass spheres acts as a static mixer. The system can be operated in an isocratic regime with programmed changes in the pressure up to 49 MPa or in the flow rate down to 1 μl/min, and in a gradient regime with an accurate gradient of 0-100% in the 10-1000 μl/min flow rate range.

Eluent delivery to open tubular columns or capillary columns with irregular packing is achieved most reliably by splitting the flow [25, 105] and sending most of the liquid through a restricted parallel channel. Clearly, the creation of this sort of flow does not guarantee the constancy of the elution rate since changes in the solvent viscosity or the accumulation of pollutants in the column will cause alterations in the splitting ratio due to changes in the hydraulic resistances. Another drawback is that the greatest advantage of capillary chromatography is its economy of the mobile phase, and this is lost if the splitting ratio is high. One of the most important hardware tasks is therefore the development of syringe pumps that can provide stable flows at rates of 0.1-1000 nl/min under pressures of up to 40 MPa.

An essential requisite of high-performance capillary liquid chromatography is the ability to use gradient elution. Only then can the column's high performances and large peak capacities be realized, i.e., multicomponent mixtures separated in reasonable periods of time. We shall now consider solvent programming methods for capillary and microbore columns.

The simplest technique was suggested in 1973 by the pioneers of microbore chromatography, Sandakhchiev et al. [8, 9], and is called the "preformed gradient" method. A syringe pump is operated so that it sucks in accurately measured aliquots of eluents with differing compositions into an intermediate capillary. Since the diffusion rate is slow, the zones with the differing compositions do not mix appreciably during the experiment. The intermediate capillary is then connected to the column, and the pump begins to discharge the eluents in the reverse order from which they were sucked in. A preformed gradient may also be created in the retention capillary by drawing up solution from a mixing reservoir in which the change in composition is accomplished in the usual way, viz., by adding the final solvent through another pump [38]. Although simple and almost equally suitable for capillary and microbore columns of any diameter, preformed gradients are subject to all the disadvantages of stopped flows (see Section 4.4), nor do they provide adequately reproducible retention volumes for component identification.

An important advance was therefore the development of automated gradient devices for microbore chromatography [12, 53, 67, 265, 266]. As in conventional high-performance liquid chromatography, two approaches are possible, i.e., mixing in the high-pressure line and mixing in the low-pressure line. In the first case, the solution is delivered using two pumps (syringe or reciprocating) into a static or dynamic mixer that is situated directly in front of the injection system. In the second case, a single reciprocating pump is used and the solvent is mixed in the intake line, the composition being controlled by a rapidly acting proportionating valve. Two-pump systems have the advantages that the actual composition more accurately adheres to the programmed one, the gradient is retained for a shorter period of time, and the eluent degassing requirement is lower. They are particularly suited to capillary chromatography in that the more suitable syringe-type pumps can be used. On the other hand, the low-pressure method is cheaper and can be used to create ternary gradients. The construction of gradient devices for conventional high-performance chromatography is discussed in several excellent reviews [267, 268], while the theory of gradient elution and the analysis of gradient distortions is covered by Quarry et al. [269].

Gradient creation for microbore columns may be automated for either of the approaches described. The single-pump system with a proportioning valve was applied by LKB Produkter AB [266], but so far its operation in the difficult gradient regimes, i.e., 0-10% and 90-100% of solvent B with supersensitive detection, has not been well investigated. The more widely utilized method is the two-pump system with mixing in the high-pressure line [53, 67, 158, 270]. The method has been proved on microbore columns [67, 270] for a variety of systems, viz., using syringe or reciprocating pumps, without a mixer, and with a dynamic or static mixer.

Any gradient system will cause systematic distortions to some degree, e.g., gradient delay or band spreading due to the finite volume of the system, as well as random distortions related to errors in the flow rate or solvent proportioning. The random distortions are the more important in that they reduce the reproducibility of the elution volume and the areas of the peaks and decrease the accuracy and sensitivity of the analysis by increasing the baseline noise. The best way of minimizing these effects is to use an effective mixer [67, 133, 270]. Scott and Kucera [53] have described a simple T-junction mixer ($V_{mix} \approx 1.5$ μl) which allows a very small amount of systematic gradient distortion but, as Schwartz et al. [67] have shown, only poorly smooths out random fluctuations. The T-junction is the best one for a mixer, and in order to create local turbulence in the channel the inlet diameters must be significantly smaller than the outlet ones. Static mixers (capillary tubes) are more effective the larger their volumes and, surprisingly, the larger the diameter of the capillary [270]. The best static mixers for capillary liquid chromatography appear to be those based on zigzag tubes [178], which are made by squeezing stainless steel capillaries between toothed rollers.

Several workers [67, 133, 270] have described dynamic mixers with magnetic armatures, one such being shown in Fig. 4.30. Dynamic mixers are good at smoothing out random fluctuations, but they significantly distort the form of the gradient, especially at the ends [67, 270]. The volume of a dynamic mixer V_{mix} must be small, but large enough to dampen out the pulsations. When the pump is a reciprocating one, V_{mix} must exceed $2V_{stroke}$, where V_{stroke} is the volume of liquid delivered by a piston stroke. Since the minimum volume per stroke of a modern pump is 30 μl, it is obvious that even for 1-mm columns the gradient generated by a reciprocating pump is inevitably far from ideal. This is true for single-pump systems too, the volume of the mixer having to be more than twice the volume of an on-off cycle of the valve [270], i.e., 90 μl for the system in [266].

Nonpulsating syringe pumps significantly reduce the demands on the volume of the mixer [270]. However, it should be remembered that the smaller the volume of the syringe, the more the operating frequency of the stepper motor can be increased for the same total delivery rate, which means that the gradient will be more accurate. For example, by modifying the electronics of the Varian 8500 syringe pump, which is far from ideal for microbore chromatography, Powley et al. [270] were only able to get 30 flow rates between 0 and 50 μl/min for frequencies below 3 MHz. This does not even compare with the Brownlee MPLC Micropump system that we described earlier. Obviously, this affected the results, which were even somewhat worse (in terms of the reproducibility of the retention time) than those obtained with the Waters M6000A reciprocating pump system [67] under similar conditions.

Unfortunately, no one has yet created an automated gradient high-pressure system using a syringe pump which can be used with packed ultramicrobore columns (d_c = 0.2-0.3 mm) or open tubular columns. When these columns are in use, the most common technique is that of splitting the flow from a commercial gradient device [25, 105] or creating an exponential gradient by using a syringe pump to deliver the final solvent into a mixing chamber that initially contains a different eluent [106, 107].

Figure 4.31 shows a diagram of the setup for creating an exponential elution gradient [107]. The mixing chamber has a volume of 40-100 μl and is made from gastight syringes and fitted with a vibrating stirrer. The form of the gradient depends on the elution flow rate and the volume of the mixer, i.e.,

$$x = a - (a - x_0) \exp\left(-\frac{Ft}{V_{mix}}\right), \qquad (4.39)$$

where x is the concentration of the modifier at the inlet to the column, a is the modifier concentration in the syringe, x_0 is the initial concentration in the mixer, and V_{mix} is the mixer chamber's volume.

The gradient approaches linearity when the flow rate and elution time are reduced. The system has been successfully tried with a 0.25-mm fused-silica ultramicrobore column, the relative standard deviation of the retention volume being less than 1%. An improvement of the method is to fill the mixing chamber with the initial solvent via a system of rotary valves and to introduce the final solvent through the loop of a six-port valve injector [106].

4.6. TEMPERATURE GRADIENTS FOR CAPILLARY CHROMATOGRAPHY

Although temperature programming is widely used in gas chromatography, the technique has not been utilized much in high-performance liquid chromatography since it does not offer many advantages over the more universal and more efficient technique of gradient elution. Furthermore, temperature gradients in conventional high-performance columns give rise to some deleterious effects, such as efficiency losses and peak asymmetry [271]. These phenomena are related to the slow flattening of the radial temperature profile in a metal-rich column containing a large quantity of poor thermal conductors, viz., the stationary and mobile phases.

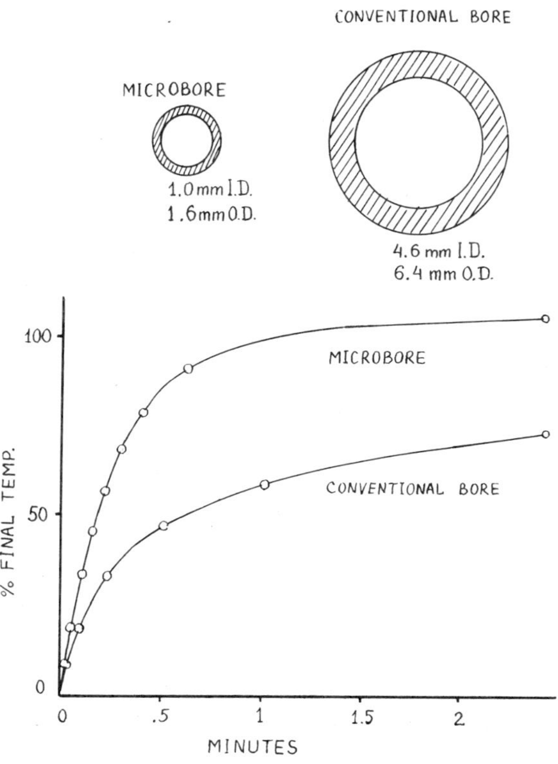

Fig. 4.32. Outlet temperature response to a step jump in temperature from 22 to 110°C for conventional columns (d_c = 4.6 mm) and microbore columns (d_c = 1 mm). (Reproduced from [62] with the permission of Preston Publications, Inc.)

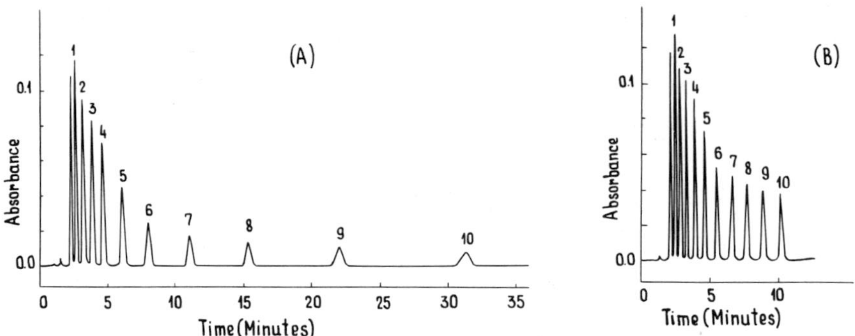

Fig. 4.33. Separation of n-alkylbenzenes A) in an isothermal regime at 25°C and B) with temperature programming from 25 to 100°C at 7.5 °C/min. A 50 cm × 1 mm microbore column packed with RP-18 (d_p = 7 μm) was used with a mobile phase of an 80:20 acetonitrile–water mixture at 250 μl/min. The pressure was 21 MPa in (A) and varied from 21 to 10 MPa in (B). UV detection at 210 nm. The peak numbers are the numbers of carbon atoms in the alkyl skeleton. (Reproduced from [64] with the permission of Elsevier Science Publishers.)

We should now perhaps reevaluate this half-forgotten technique in light of the development of microbore and capillary columns. A series of papers [51, 62, 64] convincingly demonstrating the abilities of temperature gradients for capillary liquid chromatography support this belief. We pointed out in the last section that reliable automated systems for gradient elution are not yet available for ultramicrobore packed columns and open tubular columns. Temperature gradients are therefore a viable alternative.

There is no need here to describe a temperature gradient controller for microbore columns because they are well developed for gas chromatography. It is, however, worth mentioning that stainless steel packed microbore columns can be heated under control to 100°C simply by passing an electric current through the column [64]. The advantages of microbore columns for temperature programming were demonstrated by Bowermaster and McNair [62, 64] (Fig. 4.32). For instance, a 90°C step increase in temperature took only 30 sec to become established throughout a 1-mm microbore column, whereas a 4.6-mm-diameter conventional column would require 7.8 min. The effectiveness of temperature gradients in reversed-phase chromatography is shown in Fig. 4.33, the object being n-alkylbenzenes. Compared to an isothermal regime, a temperature gradient method with temperatures from 25 to 100°C reduced the analysis time threefold, increased the detection sensitivity of the final peaks, and halved the final pressure at the inlet to the column.

Increasing the temperature reduces the viscosity of the solvent, and hence the band spreading due to mass-transfer properties. An important advantage of temperature programming over gradient elution is the constancy of the baseline, which in turn reduces the requirements on the purity of the solvents. The reproducibility of the retention volumes, $\sigma_{rel} \leqq 2.6\%$, is entirely satisfactory.

The column temperature is particularly important for open tubular columns because an effective way of overcoming their small adsorptive capacities appears to be to decrease the column temperature, which raises the distribution coefficients [79, 272].

Chapter 5

PREPARING AND TESTING CAPILLARY COLUMNS FOR LIQUID CHROMATOGRAPHY

We can assign a capillary column to one of three groups, depending on its internal diameter, preparation method, the HETP versus velocity curve, and the volume ratio of the mobile to stationary phase.

1) Packed microbore columns (PMC) with regular packing have internal diameters of 0.5-1 mm and are packed with sorbent grains that have been narrowly fractionated to between 3 and 10 μm. This is simply the capillary variant of conventional high-performance liquid chromatography columns. The category also includes packed ultramicrobore columns (PUC), which are 0.1-0.3 mm in diameter [17-19, 45-47, 273] and are usually made from glass or flexible fused-silica capillaries. When the mean particle diameter of the stationary phase lies between 3 and 10 μm, the packing in this sort of column is regular, i.e., $d_c/d_p > 10$.

2) Open tubular columns (OTC) are the analogs of the open tubular columns suggested for use in gas chromatography by Golay [151]. A number of theoretical papers [22, 23, 94, 259, 260] have shown that the optimum diameters for this type of column are between 2 and 10 μm, depending on the analysis time and pressure. In practice, however, the only high-performance columns that have so far been reliably prepared have diameters no less than 20 μm [78, 83, 87, 92]. The thickness of the stationary phase on the inner surface depends on the preparation technique and on the coating's chemical structure; it is usually between 0.2 and 5 μm thick [78, 80, 92]. Coating the surface of an open tubular column is itself an art, and some very elegant approaches have been developed for the more commonly used stationary phases. We shall consider these in Section 5.3.

TABLE 5.1. Three Column Types Used in Capillary Liquid Chromatography

Column type	d_c, µm	Permeability, K_0	Sample capacity, ng	V_s/V_m†	h_0	E_{min}‡	F, µl/min	Material
1 PMC	500-1000		10^4-10^5	1.5	2-3	2000	10-200	Stainless steel, glass, PTFE
PUC	200-300	0.001-0.002	10^3-10^4	1.5	2-3	2000	1-20	Glass, fused silica
2 OTC	10-50	0.03	2-200	10^{-2}-10^{-4}	0.3-0.8	20	0.01-1	Glass, fused silica
3 PCC*	60-100	0.01-0.06	100-1000	0.2	3.3-7	920	0.2-5	Glass

*All the data for packed capillary columns are taken from the experimental results of McGuffin and Novotny [28].

†V_s/V_m is the ratio of the volume of the stationary phase to that of the mobile phase without allowing for the internal porosity of the sorbent.

‡$E_{min} = h_0^2/K_0$ is the minimum separation impedance, according to Bristow and Knox [150].

3) Packed capillary columns (PCC) with irregular packing are also analogs of a type of column used in gas chromatography [274]. They are not popular in either gas or liquid chromatography, and only one research group [24-28] has advocated them for liquid chromatography. They are intermediate between packed and open tubular columns in terms of permeability, capacity, efficiency, and separation impedance.

Typical values for the basic parameters of each of these column types are given in Table 5.1.

In the remainder of this chapter we shall discuss the problems of preparing efficient columns of all three types, the techniques and results of testing them and correcting for the extracolumn spreading, and the possible advances in their construction technology.

5.1. EXPERIMENTAL METHODS FOR DETERMINING THE PEAK VARIANCE ASSOCIATED WITH EXTRACOLUMN BAND SPREADING

Quantitatively estimating the instrumental zone spreading is crucial in practice since the performance of each type of column may only be correctly assessed if the extracolumn spreading is accounted for. This is also true when the compatibility of ancillary equipment and microbore columns [see Eq. (3.25)] is being ascertained.

There are several experimental methods for determining the extracolumn peak variance: 1) directly measuring the variance for the analytical system without the chromatograph column [275, 276]; 2) measuring the variance for columns of various lengths and extrapolating to a zero length column [42]; 3) calculating the contribution of the extracolumn spreading to the total peak variance using the $\sigma^2_{V,i}$ versus $V^2_{R,i}$ relationship derived from the experimental chromatogram of a multicomponent mixture [165]; and 4) substituting the column by a capillary and measuring the variance for capillaries of different lengths [61].

The first method involves the withdrawal of the column from the analytical system and replacing it with a zero dead volume fitting which connects the sampling system directly to the detecting cell. The method yields reasonably accurate results, but it does not allow for the spreading in the column's filters or in the connectors. This sort of experiment, however, is very rapid and the time taken to record the concentration profile is exceedingly short. It is therefore necessary that the detector–amplifier–recorder system be rapid and it must have a time constant of less than 50 msec. Since a modern capillary liquid chromatograph (e.g., ISCO's μ-LC-system A) has its column inserted directly into the injector and detector cell, a standard zero dead volume fitting cannot be used. However, the error in the variance measurement is not very large if the injector and detector cell are connected together with a short length of fused-silica capillary (20-60 μm in diameter) inserted into the PTFE tubing with standard cone ferrules for sealing.

When samples are injected using a loop valve, the spreading in the injector system may depend on the time it takes to turn the valve and the associated pressure jumps. When the column is present, the pressure in the sampler is considerably more than when it is absent. Hence, it is doubtful whether, using this method, the injector's variance contribution $\sigma^2_{V,s}$ can be correctly accounted for. Furthermore, the chromatograph column may itself significantly reduce the real value of the sampler's spreading because of the sample concentration that occurs at the top of the column when the sample is dissolved in a solvent that promotes sorption.

The second method involves columns with various lengths, but which have the same internal diameter, are packed in the same way, and are fitted with the same filters and connectors. It is then assumed that $N_c = L/H_c$, where H_c is independent of the length of the column. If this is in fact true, then the column's contribution to the peak variance is proportional to its length, viz., $\sigma^2_{V,c} = aL$. Thus, the apparent variance that is measurable in a chromatogram, σ^2_V, is linearly dependent on the length of the column, i.e.,

$$\sigma^2_V = aL + \sigma^2_{V,ex}. \tag{5.1}$$

Above $\sigma^2_{V,c}$ and H_c are the volumetric variance and height of a theoretical plate in the absence of extracolumn spreading, and a is a constant. Thus, the extracolumn variance $\sigma^2_{V,ex}$ can be found by extrapolating the relationship back to $L = 0$. The weakness in the method is the assumption that the column's peak variance is proportional to L. It has been shown [14, 277] that there is a limiting length L_{lim} for which the assumption is valid, and that this value depends on the sorbent grain diameter and the way the column is packed. For example, for a sorbent with $d_p = 10$ μm we get an L_{lim} of 50-100 cm, while for a d_p of 5 μm we get $L_{lim} = 25$ cm. When the length of the column exceeds the limiting value, the packing in the top of the column may be packed inefficiently, and so the assumption in (5.1) is invalid.

The third method of estimating the extracolumn variance is based on the following argument. Given a constant elution rate and that the diffusivities of the species being separated are independent of k', then the height equivalent to a theoretical plate as a function of the capacity factor k' can be written [146]

$$H_i = H_a + \frac{(k_i')^2}{(1 + k_i')^2} H_b + \frac{k_i'}{(1 + k_i')^2} H_c, \tag{5.2}$$

where H_a is the contribution of molecular diffusion, eddy diffusion, and convection to the theoretical plate height, H_b is the contribution of diffusional mass transfer, and H_c is the contribution of the diffusion into a sorbent grain. For most types of chromatography H_i is only very weakly dependent on k_i' [146] because the first term in (5.2) dominates, while the sum of the second and third terms is not very sensitive to changes in k'. Hence we can write

$$H_i = \bar{\beta}_i H_0, \tag{5.3}$$

where H_0 is a constant for a given column and elution rate, and $\bar{\beta}_i$ = 1 for most species. The apparent peak variance must therefore be linearly dependent on $V_{R,i}^2$, i.e.,

$$\sigma_{V,i}^2 = \frac{H_0}{L} \bar{\beta}_i = V_{R,i}^2 + \sigma_{V,ex}^2. \tag{5.4}$$

It can be seen from Fig. 5.1, which was obtained by the hydrophobic chromatography of nitrophenols, nitrotoluenes, and nitrobenzenes on Li Chrosorb RP-18, that (5.4) is obeyed quite well, and that the intercepts, which yield the extracolumn variance, are the same for various column lengths. Moreover, the extracolumn peak variances grow as the elution rate increases, as predicted by Sternberg's theory [156]. It must be noted that the compounds used when applying this estimation technique must have similar molecular dimensions in order for diffusivities to be similar and must be chemically similar in order to exclude peak broadening due to energetic differences at the various adsorbing centers.

Compounds for which there are spreading factors not accounted for in (5.2) produce values of $\bar{\beta}_i$ greater than 1 and hence should not be used to construct the curves given by (5.4).

The relationship between $\sigma_{V,i}^2$ and k' may also be used to estimate the extracolumn spreading, but on another basis. If, as before, we assume that N_c is independent of k', then

$$\sigma_{V,i}^2 = \frac{V_0^2(1 + k')^2}{N_c} + \sigma_{V,ex}^2, \tag{5.5}$$

where V_0 is the free volume of the column.

As k' increases, $\sigma_{V,i}^2$ tends to $\sigma_{V,i}^2 = V_0^2(1 + k')^2/N_c$, and hence

$$\sigma_{V,ex}^2 = \sigma_{V,i}^2 - \lim_{k' \to \infty} \sigma_{V,i}^2. \tag{5.6}$$

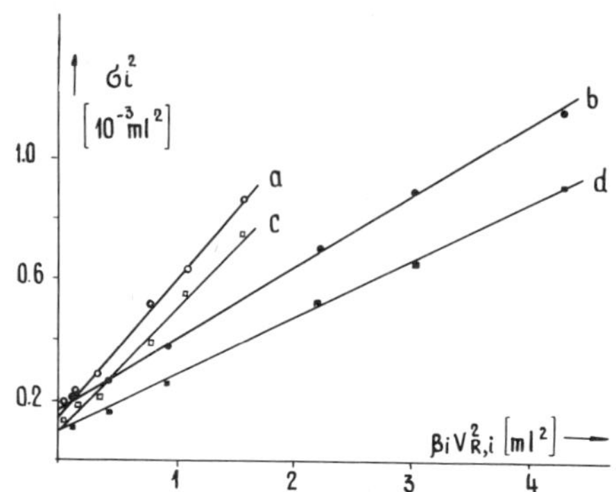

Fig. 5.1. Using the relationship between $\sigma^2_{V,i}$ and $V^2_{R,i}$ to estimate the extracolumn variance. The column's internal diameter was 1.1 mm, and it was filled with Li Chrosorb RP-18 (d = 7.7 μm). a) L = 150 mm, F = 190 μl/min; b) L = 250 mm, F = 190 μl/min; c) L = 150 mm, F = 90 μl/min. (Reproduced from [165] with the permission of Elsevier Science Publishers.)

Note that the values of $\sigma^2_{V,ex}$ that are obtained from (5.6) and (5.4) are the same for compounds with different k'. If the species being separated are very different in terms of their molecular masses, and hence their diffusion coefficients, then this method is not valid for determining $\sigma^2_{V,ex}$.

The most reliable measurements of the extracolumn variance have been obtained using the fourth method. The outlet of the injector is connected via a section of capillary to the detector, and the linear relationship between the extracolumn variance and the length of the capillary is found by carrying out the test with various lengths of capillary. This relationship is then extrapolated back to a zero length capillary to obtain the result. The method is based on Taylor's equation [152], viz., (3.51), which has been repeatedly verified. The equation shows that the volumetric variance for spreading in a capillary is proportional to its length when the flow is laminar. The choice of capillary dimensions should be guided by the limits set by Atwood and Golay [177], who established a ratio between the permissible volume of the capillary V_{cap} and the variance of a zone, i.e.,

$$V^2_{cap}/\sigma^2_{V,cap} \geqq 30. \quad (5.7)$$

If this ratio is not obeyed, then Taylor's equation is no longer valid and the concentration profile may become bimodal.

We must emphasize that whichever of the above methods is used correct results will only be obtained by determining the variance as the second statistical moment M_2 of the concentration profile [155, 275],

$$M_2 = \sigma_V^2 = \frac{\int_{V_1}^{V_2} (V - \bar{M}_1)^2 C(V)dV}{\int_{V_1}^{V_2} C(V)dV}, \tag{5.8}$$

where V is the volume of effluent, M_1 is the first statistical moment (the centroid of the peak), C(V) is the concentration profile, and V_1 and V_2 are arbitrary volumes before and after the peak for which C(V) = 0.

It has been firmly established [275] that graphical determinations of the standard deviation, either using the tangents to the inflection points or by measuring the width of the peak at $0.607C_{max}$, yield lower values of the standard deviation when the shape of the peak is not Gaussian. The concentration profile generated in the ancillary system has an exponential tail and is very asymmetric. Thus, graphically estimating the extracolumn variance not only introduces errors, but it also undermines the basis of measuring the value, i.e., the rule of the additivity of variances.

Estimating the contributions of the separate components of a chromatograph system, i.e., injector, detector cell, and connections, to the variance is an extremely difficult experimental task. The strategy developed by Hupe et al. [275] is a possible method (see Table 5.2). The variances of the concentration profiles from three systems are measured. It is essential that the extra flow cell is exactly the same as the one in the chromatograph. Then, because the variances are additive, the variances of the separate components can be calculated from the three measured values, using the formulas given in the table.

Figure 5.2 shows the relationship between the flow rate and $\sigma^2_{V,ex}$ for systems with one and two identical flow cells, and also shows the derived relationship between the flow rate and the cell's contribution to the overall variance. It is significant that the extracolumn variance as a function of the flow rate levels out for large flow rates and that the limiting extracolumn volumetric standard deviation is equal to the total volume of the extracolumn system. This finding is repeated for a variety of other extracolumn systems (see Fig. 5.3) [276]. Thus, the way the extracolumn variance depends on the flow rate characterizes the performance of the chromatograph as obviously as the h(ν) curve defines the performance of the chromatograph's column. At large flow rates a good extracolumn system behaves like an ideal mixer, i.e., $\lim_{F\to\infty} \sigma^2_{V,ex} = V^2_{ex}$.

TABLE 5.2. Determining the Variance for Each of the Components of a Liquid Chromatograph

System	Elements and element sequence in system*	How $\sigma^2_{V,i}$ was determined	$\sigma^2_{V,i}$
1	I—CC—DC	directly measured	$\sigma^2_{V,1}$
2	I—CC—CC—DC	directly measured	$\sigma^2_{V,2}$
3	I—CC—DC—CC—DC	directly measured	$\sigma^2_{V,3}$
4	I—DC	calculated	$\sigma^2_{V,4} = 2\sigma^2_{V,1} - \sigma^2_{V,2}$
5	CC	calculated	$\sigma^2_{V,5} = \sigma^2_{V,2} - \sigma^2_{V,1}$
6	DC	calculated	$\sigma^2_{V,6} = \sigma^2_{V,3} - \sigma^2_{V,2}$
7	I	calculated	$\sigma^2_{V,7} = 2\sigma^2_{V,1} - \sigma^2_{V,3}$

*I) Injector; CC) calibrated capillary, whose length and inside diameter are known and for which (5.7) is obeyed; DC) detector cell. The components are connected with zero volume fittings.

If there is a stagnant zone in any of the components of the chromatograph, then the slow diffusion of a species from the dead zone into the heart of the flow will stretch out the trailing boundary of the peak, increasing the variance in proportion to F^2/D_m [156]. Inertia in the registration system will act in the same direction, i.e., $\sigma^2_{V(\tau)} = \tau^2F^2$. In this case, which is typical of unsuccessful chromatographs, the extracolumn variance, as a function of flow rate, does not level out but worsens as the flow rate is increased. Clearly, the signs of a good chromatograph are a small total extracolumn volume $V_{ex} = \lim_{F\to\infty} \sigma^2_{V,ex}$ and a small slope on the $\sigma^2_{V,ex}$ versus F graph for small flow rates.

A good, convenient criterion for the quality of a chromatograph system as regards the extracolumn system is for the apparent efficiency N to be independent of the capacity factor k' [16]. The apparent efficiency of a chromatograph which has significant extracolumn spreading falls rapidly for small k' (Fig. 5.4A). If, however, the contribution of the extracolumn system to the overall variance is small, then the efficiency will either be independent of k' or will rise insignificantly as k' is reduced (Fig. 5.4B).

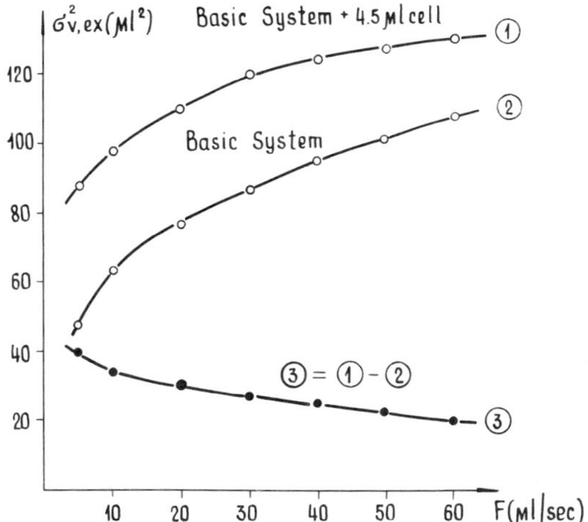

Fig. 5.2. Determining the variance of the extracolumn spreading by directly injecting the sample into the extracolumn system. The basic system consists of an injector, V_S = 1 μl, a 330 × 0.12 mm capillary, and a detector cell, V_d = 4.5 μl. (Reproduced from [275] with the permission of Elsevier Science Publishers.)

5.2. PACKED MICROBORE COLUMNS. RELATIONSHIP BETWEEN PERFORMANCE AND PACKING METHOD

5.2.1. Influence of Column Diameter on the Height of a Theoretical Plate. Wall Effect

The influence of column diameter d_c on the height equivalent to a theoretical plate (HETP) has been poorly studied experimentally and is one of the most intricate topics in chromatography theory. Although the exact nature of the $H(d_c)$ relationship has not been established, the problem must be analyzed, albeit empirically. It was not long ago that the incorrect impression that the column diameter could not be reduced below a certain limit due to wall effects [147] was widespread. The delusion was bolstered by the publication of experimental results which indicated that the efficiencies of microbore columns were low [278]. Even though these observations were due to nothing more than a large amount of extracolumn spreading and the inability of the investigators to pack the small columns effectively, they were misconstrued as being due to the fatal role of the wall effect in microbore columns and served to compromise the idea of microbore chromatography. There is no need to reestablish the idea now that Scott and Kucera [14, 52] have overwhelmingly demonstrated that well-packed, stainless steel 1-mm microbore columns can achieve efficiencies close to the theoretical

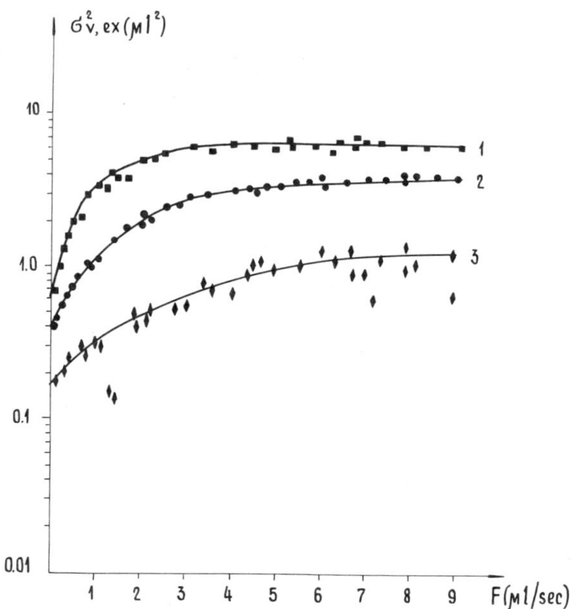

Fig. 5.3. Extracolumn variance versus flow rate for a microbore chromatograph system. 1) V_s = 0.5 μl, V_d = 1 μl, connecting capillary 155 × 0.25 mm; 2) V_s << 0.5 μl (split flow), V_d = 1 μl, connecting capillary 115 × 0.25 mm; 3) V_s << 0.5 μl, V_d = 1 μl, no capillary. (Reproduced from [276] with the permission of F. Vieweg & Sohn.)

limit for liquid chromatography, i.e., $H_{min} = 2d_p$, results that would be exceedingly good for conventional columns. Outstandingly good results, for which $2d_p < H_{min} < 3d_p$, have also been obtained for glass and fused-silica ultramicrobore columns [17-19, 273].

We shall try to throw light on the role that the wall effect, which is associated with the lower density of packing near the walls as compared with the fairly uniform packing in the center of the column, does in fact play in microbore columns. The wall effect is the local increase in the flow rate that occurs within a few sorbent particle diameters from the column walls and which distorts the eluent flow profile across the column from what it would be, given an isotropic medium. The radial mass transfer due to eddy diffusion in a column packed with sorbent considerably flattens the parabolic profile that is typical of the flow profile of fluid in a tube. However, the anomalous permeability of the packing to the liquid within a few sorbent grain diameters from the wall means that the flow velocity near the wall passes through a maximum before dropping to zero at the wall itself [280]. The column cross section can be divided into two regions, viz., an iso-

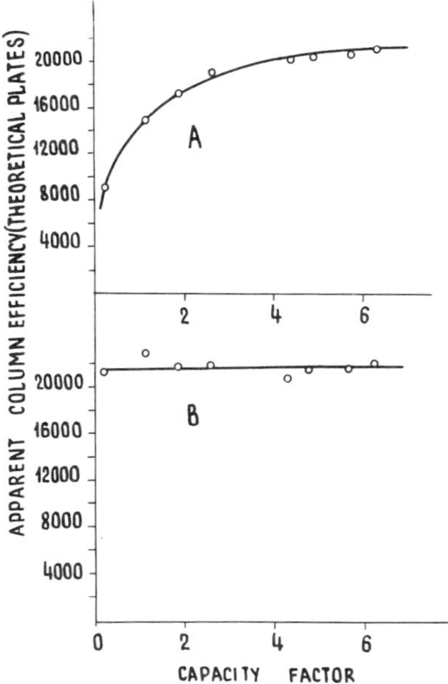

Fig. 5.4. The efficiency versus the capacity factor for a microbore column with A) a large and B) a small amount of extracolumn spreading. The 500 × 1 mm column was filled with Zorbax ODS (d_p = 8 μm) and with a flow rate F of 10 μl × min^{-1}. (Reproduced from [16] with the permission of Elsevier Science Publishers.)

tropic central section, in which the flow velocity is approximately constant; and a layer near the wall, in which the velocity first rises rapidly and then falls to zero. The existence of a velocity profile is a cause of the extra spreading of the chromatographic zones and is why the HETP is dependent on the diameter of the column [281-285].

An exact expression of the HETP of nonsorbing species can be obtained by the traditional method of taking the Laplace transform of the differential material balance, which is

$$\frac{\partial C}{\partial t} + u(r)\frac{\partial C}{\partial z} = \frac{D_r}{R^2}\frac{1}{r}\frac{\partial}{\partial r}\left(r\frac{\partial C}{\partial r}\right) + D_z\frac{\partial^2 C}{\partial z^2}, \tag{5.9}$$

where u(r) is the velocity profile across the column, D_r and D_z are the radial and axial (longitudinal) quasidiffusional coefficients, r and z are the radial and axial coordinates, and R is the radius of the column.

However, this method is difficult because of the necessity of the knowledge of C(r, z), $D_z(r, z)$, and $D_r(r, z)$. Giddings therefore has developed [281] a semiempirical theory which establishes a relationship between the transcolumn velocity profile and the HETP. Despite its simplifications, Giddings' model takes into account the relationship between the axial dispersion due to the differences in the velocities across the column and the radial diffusion. The theory yields the following expression for the contribution of the transcolumn velocity profile to the HETP when the column is sufficiently long:

$$H_T = \frac{1}{128}\left(\frac{\Delta u}{\bar{u}}\right)^2 \frac{\bar{u}d_c^2}{D_r}, \tag{5.10}$$

where Δu is the difference between the peripheral and central velocities, and $\bar{u}$ is the average linear velocity of the eluent across the column.

Although Eq. (5.10) gives a qualitatively correct picture of the relationship between the factors affecting axial spreading (the flow profile) and those affecting the axial zone sharpening (radial diffusion), Guiochon showed [31] that it produces unreasonably high quantitative results.

In order to eliminate this inconsistency, we introduce an expression for the HETP in terms of the transcolumn profile which takes into account the extent of the wall region by using a parameter like the one suggested by Knox [147, 285]. For simplicity, we assume that there is a discrete velocity distribution across the column, i.e., $u = u_W$ for $(R - \delta_W) < r < R$ and $u = u_B$ for $0 < r < (R - \delta_W)$, where r is the radial coordinate, R is the column radius, and δ_W is the thickness of the wall region. The ratio between the velocity in the wall region and that in the main, central part of the column, $u_W/u_B = \kappa$, is determined by the packing density in the two regions. The weighted average velocity of the mobile phase can be found thus

$$\bar{u} = f_W u_W + f_B u_B, \tag{5.11}$$

where f_W is the extent of the wall region, and $f_B = 1 - f_W$ is the extent of the central region.

Clearly,

$$f_W = \frac{\delta_W}{R}\left(2 - \frac{\delta_W}{R}\right), \tag{5.12}$$

$$f_B = \left(1 - \frac{\delta_W}{R}\right)^2. \tag{5.13}$$

The low variance within the axial distance of a step of a random walk in the flow direction, σ_z^2, can be defined as the variance of a discrete distribution, i.e.,

$$\sigma_z^2 = f_w(u_w - \bar{u})^2(\Delta\bar{t})^2 + f_B(u_B - \bar{u})^2(\Delta\bar{t})^2 =$$
$$f_w(1 - f_w)(u_w - u_B)^2(\Delta\bar{t})^2 =$$
$$\frac{f_w(1 - f_w)(\kappa - 1)^2}{[f_w(\kappa - 1) + 1]^2}(\Delta\bar{t})^2\bar{u}^2, \tag{5.14}$$

where $\Delta\bar{t}$ is the average time for a step of a random walk. When f_w is much smaller than unity, (5.14) simplifies to

$$\sigma_z^2 = \frac{\delta_w}{R}(\kappa - 1)^2(\Delta\bar{t})^2\bar{u}^2. \tag{5.15}$$

Radial diffusion tries to equalize the species concentration between the wall region and the main column region, acting so as to reduce the axial spreading. The average time for the species to be transferred from the wall layer can be determined using Einstein's equation

$$\Delta\bar{t} = \left(\frac{\delta_w}{R}\right)^2 \frac{R^2}{8D_r}. \tag{5.16}$$

Thus, the number of steps is equal to the number of mass-exchange events between the wall and the central regions, i.e.,

$$\bar{n}_{wB} = \frac{t_0}{(\Delta\bar{t})} = \frac{8LD_r}{\bar{u}(\delta_w/R)^2R^2}. \tag{5.17}$$

An expression for the HETP in terms of the transcolumn velocity profile follows from the above, i.e.,

$$H_T = \frac{\sigma_z^2 n_{wB}}{L} = \frac{(\delta_w/R)^3(\kappa - 1)^2\bar{u}R^2}{8D_r}. \tag{5.18}$$

The following is valid for D_r:

$$D_r = \gamma D_m + \lambda_r d_p \bar{u}, \tag{5.19}$$

where λ_r is the radial eddy diffusivity and is equal to 0.08 [282, 285], while γ is a factor due to the convolutions of the interparticle channels and is equal to 0.7 [285].

Using (5.18) and (5.19) and transforming to reduced parameters, we get

$$h_T = \frac{(\delta_w/R)^3(\kappa - 1)^2(d_c/d_p)^2}{22.4\nu + 2.56}. \tag{5.20}$$

This equation, in contrast to (5.10), yields reasonable results for the reduced HETP for $0 < \delta_w/R < 0.3$ and $1 < \kappa < 4$. Remember, however, that the equation is only strictly valid if $\delta_w/R \ll 1$.

TABLE 5.3. Properties of Bodies Obtained by Packing Spheres of Radius R_p [286]

Type of packing	Coordination number n_c	Interparticle porosity ε_0	Radius of sphere inscribed in a cavity
Dense hexagonal	12	0.26	$0.225R_p$
Centered tetragonal	10	0.302	$0.291R_p$
Simple hexagonal	8	0.395	$0.527R_p$
Simple cubic	6	0.476	$0.732R_p$
Tetragonal	4	0.666	$1.0R_p$

TABLE 5.4. Limiting Dimensions of the Wall Region for Well-Packed Chromatograph Columns, $h_T = 1$ and $\kappa = 2$

d_c, mm	4		1		0.2	
d_p, μm	5	10	5	10	5	10
δ_w/R	0.025	0.04	0.063	0.1	0.18	0.29
δ_w, μm	50	80	32	50	19	29
δ_w/d_p	10	8	6.3	5	3.7	2.9

The possible values of $\kappa = u_w/u_B$ can be estimated from the data in Table 5.3. We know [154] that the interparticle porosities of practically realizable packings lie in the range $0.36 < \varepsilon_0 < 0.485$, i.e., they are intermediate between simple cubic and simple hexagonal packing. If we assume that the packing in the central part of a packed column has a coordination number of eight and that in the wall layer a coordination number in the range $4 \leqq n_c \leqq 6$, then the ratio between the linear eluent velocities in the two regions is $\kappa = u_w/u_B = (R_w/R_B)^2$, where R_w and R_B are the respective radii of the interparticle channels, as given in Table 5.3.

In this way κ may lie in the interval $1.9 < \kappa < 3.6$. Using this datum and (5.20), we can calculate the possible dimensions of the wall region corresponding to $h_T = 1$ and $\nu = 3$, i.e., properties of reasonably well-packed columns. The results are given in Table 5.4.

Knox, Laird, and Raven [285] have determined a parameter range for the number of particle diameters in the wall region using delicate experiments on 11.7-mm columns filled with impermeable glass spheres ($d_p = 63$ μm) for $h_{min} = 2$, viz., $15 < \delta_w/d_p < 30$. But if

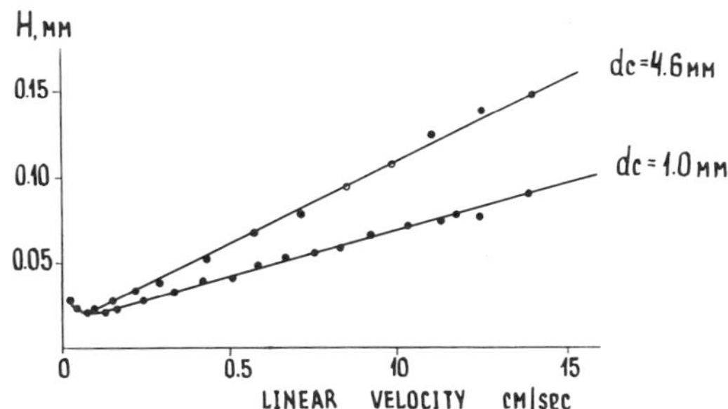

Fig. 5.5. HETP versus elution velocity for columns of differing diameters. The column length was 50 cm, the stationary phase was Partisil, and d_p was 10 μm. (Reproduced from [14] with the permission of Elsevier Science Publishers.)

we want $h_T = 1$ for $\nu = 3$, then from (5.20) we must have $\delta_w/d_p = 6.1$. This indicates that even in what we now consider to be high-performance columns ($h_{min} = 2$), the transcolumn velocity profile makes a large contribution to the spreading of the chromatographic zones, and hence there is considerable scope for enhancement, possibly to $h_{min} < 1$, by improving packing techniques so as to reduce the width of the wall layer.

It is now easy to explain, on the basis of (5.20), the contradictory experimental results concerning the influence of column diameter on the HETP. Clearly, if in packing columns of differing diameters it is possible to keep the value of δ_w/R constant, then reducing the column diameter will reduce the HETP, in this case h_T being proportional to d_c^2. This is apparently what Kucera [14] was able to do in his experiments, as shown in Fig. 5.5.

On the other hand, if the absolute value of δ_w/R is maintained for columns of different diameters, then h_T will be proportional to $1/d_c$ and the column performance will fall off as the diameter is decreased. Finally, if neither δ_w nor δ_w/R are kept constant, then a relationship with a minimum point will be found for h_T and d_c, as for example in Fig. 5.6 [17], which shows Yang's experimental results for ultramicrobore fused-silica columns.

The transcolumn velocity profile is also associated with the effect observed by Knox and Parcher [284], i.e., the effect of the infinite-diameter column. The incorrect interpretation of this effect also compromised the idea of microbore liquid chromatography. Knox and Parcher noticed that if a sample was injected into the center of a column that was wide enough for the sample to pass through the column without it being transported to the wall region by radial

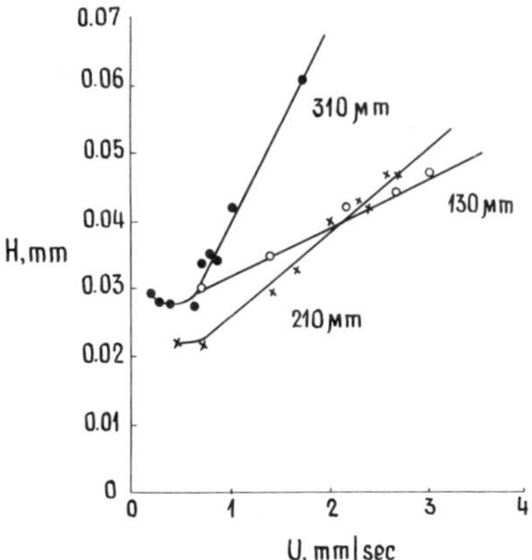

Fig. 5.6. HETP versus elution velocity for fused-silica ultramicrobore columns of differing diameters. k' = 5.7, the stationary phase was Micropak MCH, and d_p was 5 μm. (Reproduced from [17] with the permission of Elsevier Science Publishers.)

diffusion, then the HETP was reduced. The critical parameter for this effect is R/σ_R, where σ_R is the standard deviation (length units) for the radial diffusion lengths and $\sigma_R = (1.4/\nu + 0.16)d_p u$ [147, 285]. If $R/\sigma_R \approx 0.6$, then 95% of the sample reaches the column wall, while if it is greater than 2, then only 5% of the sample reaches the wall.

The infinite-diameter phenomenon is important if δ_W is constant and independent of R. If, however, the column packing makes δ_W fall off uniformly with R, then it becomes less necessary to introduce the sample into the center of the column because the acceleration of radial mass transfer, which occurs in microbore columns, becomes an advantageous factor in itself [cf. (5.20)].

The theoretical considerations we have presented in this section demonstrate the very close relationship between the packing regime and the wall effect. There is significant room for improving the efficiencies of microbore columns (see Table 5.4) if the packing process can be changed so as to reduce the thickness of the wall layer to one or two times the diameter of a sorbent grain. We shall now consider the problems of packing columns.

5.2.2. Packing Microbore Columns

The skill of effectively packing a high-performance column with a stationary phase, whose sorbent particles have mean diameters between 3 and 10 μm, is at present as much an art as growing protein crystals for x-ray analysis. That the vast army of chromatographers have not been able to make sense of this problem is very sad, and unfortunately the firms that have a great deal of experience in packing and selling finished columns do not wish to reveal their know-how because the price of a packed and tested column is very many times greater than the sum of the prices of the column and the stationary phase taken separately. On the other hand, those papers that partially deal with the topic have tended to muddy rather than clarify matters. The published data are clearly insufficient for a generalization: in the worst cases descriptions of packing methods only provide values of the column efficiencies, while in the best cases the descriptions include the velocity dependence of the HETP and values of the separation impedance. The recommended procedures are often diametrically contradictory; thus, the literature includes high-viscosity [287] and low-viscosity [288, 289] methods, techniques using dilute [290, 291] and concentrated [292] slurries, and packing at moderate [293] and very high [14, 294] pressures.

Packing reversed-phase columns is particularly difficult. The solvents used to prepare the slurry that have been recommended in different papers have varied in polarity from alcohols [290, 295] to carbon tetrachloride [288]. The situation has become even more confused with the appearance of microbore columns. The first research group, under Ishii [37-39], used 0.5-mm PTFE microbore columns and packed them with pressure drops of less than 10 MPa. Because acceptable efficiencies could not be achieved with these columns (the best result being $h_{min} \geqq 6$), Scott and Kucera [14-16, 52-54] used 1-mm steel columns instead and found that pressures of 175 MPa were needed to get as efficient a packing ($h_{min} \approx 2$) as would require 70 MPa for a conventional 4-mm column. Later papers on packed 1-mm steel microbore columns [63, 66, 296, 297] also indicated that very high pressure drops (greater than 100 MPa) were needed for packing.

It would thus seem that packing microbore columns is much more difficult than packing conventional columns. However, as soon as research groups [17-19, 46-48, 273] began working with the even smaller diameter columns (0.2-0.3 mm) made from fused silica and glass they found that they could easily get excellent results ($2 < h_{min} < 3$) with packing pressure drops of less than 45 MPa, and in some cases [18, 47, 48] the packing was compacted manually using gas-tight syringes! Moreover, while the upper length for effectively packing 1-mm steel microbore columns, with a pressure drop greater than 100 MPa, was 25 cm (d_p = 8 μm) [14], 0.2-mm silica ultramicrobore columns could be packed to a length of 1 m using a stationary phase with 3-μm particles [17, 19]!

In this section we shall try to break through the many contradictions associated with packed columns and try to draw a more or less consistent picture of the process.

When a column is being packed, both the local interparticle fluctuations in packing density (the intrachannel effect) and the column scale fluctuations (the transcolumn effect) are deleterious. If a column is packed with a slurry that is stable to aggregation, then for modern microparticle sorbents, which have very narrow particle size distributions, local density fluctuations in the packing will be absent. Obviously, the packing pressure cannot exceed the mechanical strength of the particles; otherwise, they will be crushed and the resultant structure will not be homogeneous. Note that the siliceous sorbents that have very large specific pore volumes ($\bar{V}_p \geqq 2$ mg/g) have very low mechanical strengths (e.g., Merck's Li Chrosphere Si-300 or Toyo Soda Corp.'s TSK-GEL SW-3000) and packing pressures above 15 MPa cannot be used. The suspension must also be sedimentation stable, but since the sedimentation rate of particles smaller than 10 μm is very slow, there is no need to take any special precautions such as balancing the density of the suspension [298].

The major danger is therefore irregularities in the packing across the column that are due to the wall effect, which we covered in Section 5.2.1. It may be shown that the contribution of this effect to the height equivalent to a theoretical plate, for a chromatographic column, is closely related to the packing regime of the column.

The first stage in the packing process is shown in Fig. 5.7. When particles strike a layer of settled stationary phase, they are pressed into the layer by the fluid's viscous force of friction, i.e.,

$$f = 3\pi\eta u_s d_p, \qquad (5.21)$$

where η is the viscosity of the dispersing liquid, and u_s is the flow velocity of the slurry in the unpacked space of the column.

The question then arises as to how high must the absolute pressing force be in order to obtain the most dense random packing. A solution can be obtained using the data collected by Every and Ramsay [286] about the way powdered siliceous aerogel ($d_p = 4\cdot10^{-9}$) responds to being dry-pressed under pressures of up to 1550 MPa. They managed to obtain an interparticle porosity of 0.36, which corresponds to a coordination number of about 9, using a pressure of 700 MPa. From this figure we can derive the requisite pressing force acting on a single particle, which is $f_{lim} = (\pi d_p^2/4)\Delta P = 10^{-9}$ kg.

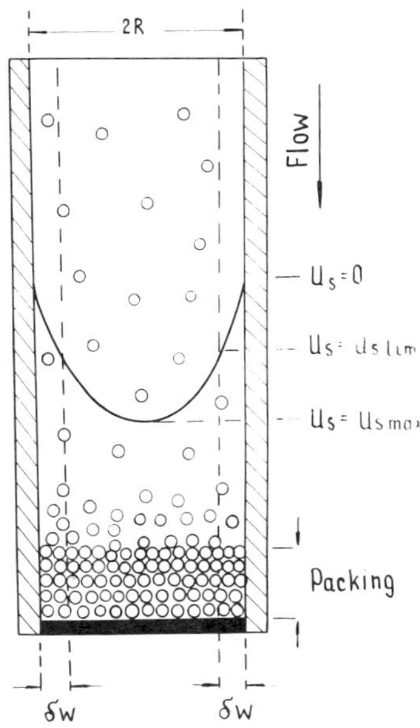

Fig. 5.7. Sketch of the first stage in packing a column by the slurry method.

Note that a porosity of 0.36 is the lowest that can be obtained in practice when packing chromatograph columns [154]. Given that f_{lim} is 10^{-9} kg, we can now estimate the minimum solvent (viscosity 10^{-3} Pa·sec) flow velocity needed for packing from (5.21), i.e., $\bar{u}_s$ = 10 cm/sec for d_p = 10 μm or $\bar{u}_s$ = 5 cm/sec for d_p = 5 μm. This rapid a flow velocity can never be achieved in practice. For instance, in Kucera's paper [14], a column was packed with a 10-μm sorbent under a constant pressure of 175 MPa, and the final flow velocity was 8 cm/sec, or 80% of the requisite amount. As the slurry flows down the tube, the flow profile is parabolic, i.e.,

$$u_s(r) = 2\bar{u}_s(1 - r^2/R^2), \tag{5.22}$$

where $\bar{u}_s = k_0 d_p^2 \varepsilon \Delta P/\eta L$ is the mean rate.

It is possible that this profile is flattened somewhat by the presence of dispersed particles, but this does not affect the point. It follows from (5.21) and (5.22) that in order to obtain a column with the desired wall-layer thickness δ_w the packing flow velocity must be

$$u_s = \frac{u_{lim}}{1 - (1 - \delta_w/R)^2}, \tag{5.23}$$

where $u_{lim} = f_{lim}/3\pi\eta d_p$.

If the packing is done at constant pressure, then the pressure drop must be

$$\Delta P = \frac{f_{lim}L}{6\pi\varepsilon k_0 d_p^3} \frac{1}{1-(1-\delta_w/R)^2}. \tag{5.24}$$

To ensure that $\delta_w/R = 0.1$, and given that $L = 25$ cm, $d_p = 10$ μm, $\varepsilon = 0.8$, and $k_0 = 10^{-3}$, then the pressure drop must be 870 MPa!

These estimates show that if the pressing force was the sole factor determining the structure of the packing, then we would never have obtained efficient columns. It was also significant that in meeting the conditions in Eqs. (5.23) and (5.24) when packing columns of differing diameters, the ratio δ_w/R will remain constant.

In practice, however, the opposite situation is also possible, as shown in Section 5.2.1; i.e., a reduction in the column diameter is accompanied by a rise in the value of δ_w/R (see Fig. 5.6). Hence the velocity at which the particles come up against a packed layer is not the only factor affecting the structure of the wall layer.

The second stage in the packing, which can be termed the packing consolidation stage, is also important. Table 5.3 shows that the coordination number in the packing can vary from 4 to 10, the most dense packing with a coordination number of 12 never being achieved in practice. After the first stage of packing, the stationary phase is a typical solid pseudoplastic; i.e., it will deform slowly if the shear stress exceeds a fluidity threshold. As long as the packing in the column is not stable, isotropic, simple hexagonal ($\varepsilon_0 = 0.39$) throughout the column, it is possible in principle to make the packing stronger and denser by inducing plastic flow. The fluidity threshold depends on the strength of the structure formed during the first stage of packing, which depends on the number of interparticle contacts and their adhesive strength. The wall layer, with its smaller coordination number, must be the first to be made denser. The calculations we made above indicate that packing consolidation must occur in any packing method; otherwise, a high-performance column would be impossible. However, a special approach may promote further positive rearrangements in the packing during the consolidation stage.

For example, the adhesive strength of the interparticle contact, and hence the fluidity threshold, may be deliberately reduced by adding a solvent that has a polarity similar to that of the stationary phase, thus promoting repulsion between the particles. The elastic solvated layer then helps the particles to slide over each other. It is also sensible, at the consolidation stage, to increase the shear stress, which can be done by displacing the slurry liquid with a more viscous solvent. Spherical particles have the least contact area and so are more easily compacted than irregularly shaped particles. The smaller the particles, the worse they pack down during the first stage of packing [see Eq. (5.24)], and the more difficult it is to redistribute them at the consolidation stage because of the larger total contact area.

It should be noted that plastic redistribution of the packing is an extremely slow process. Therefore, the simplest method to consolidate the packing is to pump the mobile phase through the packed column at the highest possible pressure for a long time (several days). At the consolidation stage the shear stress in the various density domains depends on the generalized force $\Delta P d_c^2$, i.e., it is smaller, the smaller the column diameter. Thus, microbore columns are usually more difficult to consolidate, although the roughness of the inner wall of the tube has a significant role [209-301].

During the consolidation stage, it is easier to make the packing in the wall region denser if the adhesive contacts between the packing and the walls are small. It is not surprising, therefore, to find that the ideally smooth walls of fused-silica tubes implies that the pressure drop needed to pack them is relatively low.

There is another important point regarding the consolidation stage. If there is not excess packing over the top of the column, then the packing will shrink with the formation of empty spaces in the upper part of the column, which will obviously diminish its quality.

On the other hand, if the coupling between the slurry vessel and the column is narrow, then the plastic flow of the settled adsorbent will be impeded. Hence Halász's [55] recommendation of using a precolumn during the packing is absolutely correct. Alternatively, the slurry vessel could be in the form of a cone with a smoothly changing diameter [66].

The above are all related to rigid, unswollen sorbents. There is a very simple way of consolidating columns for a "semisoft," moderately swollen gel. This is to displace the slurry solvent with one that causes the gel to swell an additional 20% or so.

The third and final stage in the ideal process of packing a column, but before it is sealed, is to fix the packing. The packing must be made as strong and stable as possible so that it does not loosen when the pressure is removed and the slurry vessel disconnected. It is especially important for nonpolar reversed-phase sorbents, since their particles become electrostatically charged by rubbing with a nonpolar slurry liquid. The most radical approach here may be to replace the solvent with one that promotes particle aggregation and removes the solvated layers. Possible solvents for reversed-phase systems are water or aqueous salt solutions.

TABLE 5.5. Basic Stages in Packing Columns and Associated Factors

<table>
<tr><th>Packing stage</th><th>Result</th><th colspan="2">Significant factors</th><th>Insignificant factors</th></tr>
<tr><td>1. Pressure-induced slurry filtration</td><td>Formation of structure</td><td>Sedimentational and aggregative stability of the slurry (solvophilicity of the stationary phase)</td><td rowspan="2">Highest possible pressure, as determined by particle crushing strength</td><td>Slurry concentration, viscosity, and column diameter</td></tr>
<tr><td>2. Packing consolidation</td><td>Reducing the strength of the packing structure and plastic flow resulting in a denser medium</td><td>Solvophilicity of the stationary phase, high solvent viscosity, sphericity of the particles, process duration, wall roughness of column, and precolumn diameters</td><td>Solvent density</td></tr>
<tr><td>3. Packing fixation</td><td>Strengthening the structure</td><td colspan="2">Solvophilicity of the stationary phase</td><td>Column diameter</td></tr>
</table>

TABLE 5.6. Characteristics of Modern Packed Columns

d_c, mm	Material	d_p, μm	L, cm	Packing pressure, MPa	h_0	ν_0	$k_0 \cdot 10^{-3}$	Coefficients in Eq. (3.5)			Refs.
								A	B	C	
1	Stainless steel	8	50	175	2.3	2	–	1.3	1.4	0.07	[16]
1	Stainless steel	8	100	180	3	1.7	–	1.4	1.7	0.19	[66]
1	Stainless steel	5	25	100	2.7	4	1.4	1.05	3.6	0.09	[296]
1.2	Polished stainless steel	8	25	60	1.8		–				[361]
1	Stainless steel	8	100	180	2.7-3.3		1.4-1.8				[297]
1	Polished stainless steel	5	13	100	2.9	2.15	1.4		–	–	[63]
0.2	Silica	5	100	42	2.52	2.5	1				[19]
0.2	Glass	10	100	50	2	2	1.5				[273]
4.2	Stainless steel	11 6.5	20	35	1.3 1.63	8 6				0.04 0.06	[303]

Only when all three stages have been completed can the pump be turned off and the column sealed by the top frit.

The strategy we have just described is given in a generalized form in Table 5.5.

The success of the packing depends on how well all three stages are completed. Clearly, the solvent must be changed at least once during the packing process because the demands on the solvent in the different stages are incompatible.

The best results achieved so far for microbore columns are given in Table 5.6. A comparison with the best results available for conventional columns is given in the last line. It can be seen that the packing procedure for microbore columns is still poorly developed, and hence there is scope for increasing their performance by two- or threefold. The path to achieving this lies in improving stages two and three.

Consider, by way of example, the packing procedure used by Welling et al. [301].

A 1.2-mm-diameter column was packed with Du Pont's alkylsilylated silica gel Zorbax BP-ODS, which has a particle size of 8 μm. The packing was done at a constant flow rate of 1.2 ml/min and a final pressure drop of 60 MPa. The slurry solvent was carbon tetrachloride ($\eta = 9.7 \cdot 10^{-4}$ Pa·sec), which was displaced by methanol ($\eta = 5.45 \cdot 10^{-4}$ Pa·sec). Over a total period of 30 min, 6.5 ml of carbon tetrachloride and about 28 ml of methanol were passed through the column.

It is immediately apparent that this is a rather short period of time and clearly insufficient for the packing to be consolidated completely. The switch to the more polar solvent, methanol, should have helped fix the packing, while its lower viscosity and lower affinity to the sorbent should have diminished the effectiveness of the consolidation process. However, the column was successful ($h_0 = 1.8$), and this shows that it should be possible to get an h_0 of about 1 if the packing regime is optimized.

Both steel and silica packed microbore columns can be connected together without any change in h_0 or k_0 to form a very long column with an overall efficiency of a million theoretical plates. Menet et al. [297] have achieved this experimentally with a compound column 22 m long and a pressure of 60 MPa, the time taken for the elution of the unretained component being 16 h.

It should be repeated that the published papers on packing microbore columns are very uninformative. We would like to recommend that the following be published in future reports to make comparisons more useful: packing pressure, overall column porosity, interparticle porosity, h_0, ν_0, the coefficients of Eq. (3.5), and a detailed description of the packing procedure. A generalization of the resultant material would demonstrate whether the idealized picture of packing described in this section is, in fact, realistic.

5.3. PREPARING AND TESTING MICROCAPILLARY COLUMNS

5.3.1. Packed Capillary Columns with Irregular Packing

This sort of column is prepared by taking a section of glass tubing, with 0.25-mm internal diameter and 5-mm-thick walls, packing it completely with sorbent (particle diameter between 10 and 30 μm), and drawing it out into a capillary of the requisite dimensions using a standard glass-drawing device, such as Shimadzu's GDM-1B.

The packing's homogeneity in the final capillary depends very much on the quality of the packing in the initial tubing. At the softening temperature of the glass the sorbent particles partially sinter and partially become embedded in the capillary glass, which stabilizes the packing. The glass used should be an easily softenable one, while the sorbent is limited to those materials which are neither decomposed chemically nor structurally changed by the temperatures needed to melt the glass. Novotny et al. have shown that several sorts of silica gel [24, 25, 28], as well as activated alumina [24], are sufficiently thermally stable.

Note that the temperatures needed to prepare the column must dehydrate the surface of silica sorbents and hence reduce their adsorption activities. The surface of the stationary phase in a finished column can be rehydrated by a long period of washing the sorbent with hot 1 M HCl.

In order to get a homogeneous packing it is essential for the sorbent to be narrowly fractionated in size. The best results have been achieved with sorbent particles 30 μm in diameter in columns with internal diameters between 50 and 80 μm. Capillaries with larger diameters have been shown to have less stable packing, and so it is partially broken up during chromatography.

Figure 5.8 shows a typical relationship between the height equivalent to a theoretical plate and the elution velocity, using data obtained by Tsuda and Novotny [24] for a capillary column packed with aluminum oxide. At the slowest elution rate that could be measured, the HETP is close to the theoretical minimum, i.e., $H = 3.6d_p$. However, the graph of the HETP versus elution rate is unusual in that, after the curve has leveled out, a new upsurge ap-

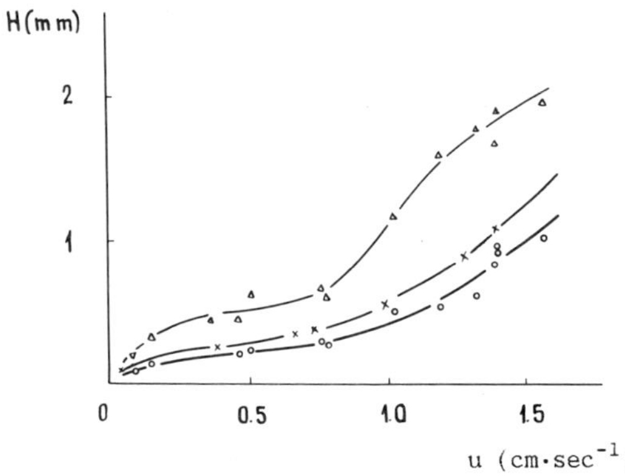

Fig. 5.8. HETP versus the elution velocity for a packed capillary column. L = 14 m; d_c = 75 μm; d_p = 30 μm. o) Benzene; ×) methyl benzoate; Δ) quinoline. (Reproduced from [24] with the permission of the American Chemical Society.)

pears. The reason for this anomaly has not been established exactly, but it is apparently due to the irregularity of the packing. This serious flaw of packed capillary columns is perhaps why they are not very popular.

The advantage of this sort of column is that it can be made as long as desired and yet the structure of the packing will be the same along the whole length. The very high interparticle porosity, which is characteristic of irregular packing, makes the chromatography very efficient at relatively low pressures. For instance, an efficiency of 85,000 theoretical plates was achieved in a capillary column 29 m long, the internal diameter being 75 μm and the particle diameter 30 μm, at a pressure of only 10.5 MPa [24]. The limitation in the choice of sorbents to thermally stable ones means that the potential of this sort of column is also limited. However, when organosilanes are used, it is possible to modify the stationary phase in situ and so obtain columns with various selectivities for reversed-phase, ion-exchange, and normal-phase chromatographies [26]. The hydroxyl groups on the surface of alumina can be reacted with organosilanes in the same way they react with the silanol groups of silica. The following have been used as modifiers: octadecyltriethoxysilane, N-(β-aminoethyl)-γ-aminopropyltriethoxysilane, and β-cyanotriethoxysilane. In order to modify the sorbent in a finished packed capillary column, hexane is passed through it for several hours at 50°C and a flow rate of 10 μl/min. Then 150 μl of a 2% solution of the corresponding silane in toluene and again the same quantity of a solution of the organosilane is fed through the column. When the reaction has finished, the column is washed for 24 h with hexane, and then the working eluent is passed through the column to stabilize the baseline.

5.3.2. Open Tubular Columns

Open tubular columns are prepared mainly from borosilicate or soda-lime glass. It is also possible to use the fused-silica capillaries made by Scientific Glass Engineering (UK), but it is difficult to make the stationary phase adhere to the inner surface of the column.

In order to produce a tolerable sorptive capacity the very small inner surface area of an open tubular column must be enlarged by creating an isotropic porous layer on the column wall. A number of skillful methods have been developed to do this on glass walls, but none of them are suitable for fused-silica surfaces. Thus, silica open tubular columns, which Yang has investigated [259, 260], have only been used by other authors when large sorptive capacities are not required, e.g., in single-phase hydrodynamic chromatography [96] or for testing new types of detectors [101-104].

The porous layer can easily be formed on the internal surface of glass capillaries by chemical [78, 82, 83, 92] or electrochemical [94] etching, or by the deposition of SiO_2 or Al_2O_3 colloidal sols [80, 82]. Ishii et al. [78, 82, 83] have obtained some very good results with normal-phase open tubular columns coated in a quasi-silica gel. The coating is formed by etching the surface of a soda-lime glass with an alkali solution of NaOH, KOH, and LiOH. By varying the concentration of the alkali, the temperature, the etching time, and the alkali cation, it is possible to alter the depth of the etching, the surface porosity, and the dimensions of the silica aggregates obtained on the surface within broad limits. The best results were obtained using 1 M NaOH at 25-30°C for 2-6 days, or at 40-54°C for 2 days. The coating obtained under the optimal conditions has a well-developed surface with silica aggregates about 1 μm wide and 0.5 μm deep.

Another method for obtaining a layer of silica gel on the inner surface of an open tubular column is to etch the glass chemically with an ammonia or tetramethylammonium hydroxide (TMA) solution [92]. The silicic acid salts formed by the etching are deposited in a thin layer on the surface of the column. These are then decomposed by heating to form SiO_2 and volatile products, such as ammonia and trimethylamine. The resultant silica particles are practically all the same size. The actual process is extremely simple. A glass capillary is half filled with reagent, closed, and kept at 150-170°C for 2-24 h. After cooling, the solution is blown out with nitrogen and the column is heated in a stream of nitrogen for 10-120 min at 150-350°C. It is possible to vary the thickness of the resultant silica layer by changing the composition of the reagent, with thicknesses from $7 \cdot 10^{-2}$ to 0.3 μm being attainable. The optimum concentration of TMA is 3-4%, while that of ammonia is 8-9%, the TMA yielding a significantly stronger and more uniform coating than the ammonia.

When a 0.5-2.7% solution of NH_4HF_2 in methanol is used for the etching, a quasi-silica gel coating is again obtained, but with a completely different surface structure [304, 305]. The structure is made up of nets of randomly distributed silica needles, or whiskers.

Jorgenson and Guthrie [94] have described an electrochemical etching treatment of capillaries made from borosilicate glass. Their apparatus is illustrated in Fig. 5.9. During the process a 0.2 M phosphate buffer (pH 7) was passed through the tube while the tube itself was immersed in a bath of hot (85°C) 3 M HCl. A dc potential of 3 kV was applied across the column walls by grounding the bath and placing the anode in the buffer in the outlet reservoir. It turned out that for the treatment to be successful the capillary had to be heat treated beforehand for 24 h at 500-600°C. The etching treatment itself is completed after 24 h with the formation of a rough inner wall, the roughness depth being $0.2\ \mu m < d_f < 0.6\ \mu m$. A 15-μm ID column was prepared, using this technique, that had a capacity factor twice that of a column with an ID of 65 μm. This indicates that a large surface area per unit volume can be achieved even in ultrasmall-diameter open columns by etching.

A porous layer can be prepared from a silica or alumina colloidal solution [80, 82]. First, the sol is made from particles of the required dimension by dispersing an aerogel using ultrasound either in water or in another solvent that prevents the particles from aggregating. The concentration of the sol is from 1 to 5%. A capil-

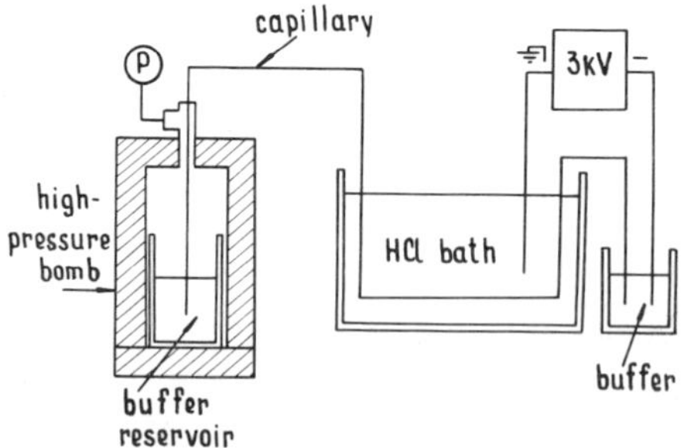

Fig. 5.9. Layout of the apparatus for electrochemically etching the inside of an open tubular column. (Reproduced from [94] with the permission of Elsevier Science Publishers.)

lary made from borosilicate glass is then treated with 1 M NaOH to roughen its inner surface, and then filled with the sol and dried in an oven, with the temperature being raised evenly from 20 to 130°C. It should be noted that using this method it is problematical to prepare open tubular columns with internal diameters smaller than 10 μm. This is because steric hindrances arise when filling this size of column with sols containing particles 100-200 nm in diameter. However, colloidal solutions with smaller particles (3-10 nm) are very unstable.

The stationary phase thus obtained can either be directly used for adsorption chromatography or physically and chemically modified for partition, ion-exchange, or reversed-phase chromatography.

In partition liquid–liquid chromatography the coating is made from Apiezon H, β,β'-oxydipropionitrile (ODPN), SE-30, or an oligo-ethylene glycol, the coating being made dynamically by the evaporation of a solution of the appropriate compound. SE-30 is delivered as a 4% solution in petroleum ether, and passed through the tube at 4 cm/sec [74, 82]. The coating is then dried in a stream of nitrogen at 230°C. It may either be applied directly to the glass or the glass can first be treated with trimethylchlorosilane to make it hydrophobic. A stable coating can only be obtained from ODPN or the other liquid phases if the stationary phase is applied onto a rough surface layer that is very adhesive, i.e., one made by etching the capillary. An even layer can be made from ODPN by passing a solution of ODPN in dichloromethane through the column for 20 min and then drying it in a stream of nitrogen at 52°C [76].

Stationary phases with different selectivities for normal-phase, reversed-phase, and ion-exchange chromatographies can be obtained in open tubular columns by chemically modifying in situ the surface of a porous layer made from either alumina or silica. The following reactions can be used for this purpose:

$$\equiv Si{-}OH + SO_2Cl_2 \rightarrow \equiv SiCl \xrightarrow{+\ RMgCl(RLi)} \equiv Si{-}R \qquad (5.25)$$

$$\equiv SiCl \xrightarrow{+\ RNH_2} \equiv Si{-}NH{-}R \qquad (5.26)$$

$$\equiv Si{-}OH + XSiR_3 \ (\text{or}\ XSiR'_2R) \longrightarrow HX\uparrow + \equiv Si{-}O{-}SiR_3 \ (\text{or}\ \equiv Si{-}O{-}SiR'_2R) \qquad (5.27)$$

where X = Cl or CH_3O, and R' is an alkyl group.

Grafting a functional group via an Si–C bond [306, 307] using (5.25) or via an Si–N–C bond [308] using (5.26) is far less popular for bonded phases than using the very well-studied reaction of (5.27) between silanol hydroxyls and organochlorosilanes or organoalkoxysilanes. A similar method may be to modify the surface of an alumina-based stationary phase, too, with aluminum oxide [26]. In reaction (5.27) silanes with two or three functional groups, instead of the one X group shown, may be used. However, it is thought that the best results are to be obtained with monofunctional silanes because if there is any trace of water, then silanes with more than one reactive group will polymerize and form crosslinked structures. Even though the coating is then denser, the pore size and volume are greatly reduced and the resistance to internal mass transfer is raised. When the silica is at its maximum hydration, the density of hydroxyl groups is 8 μmoles/m² [309]. The degree of covering can be estimated thus:

$$\alpha_{\mathrm{cov}} = W/M\bar{S}_{\mathrm{cor}}\ (\mathrm{moles/m^2}), \qquad (5.28)$$

where W is the weight of grafted functional groups per gram of stationary phase, M is the molecular mass of the grafted groups, and $\bar{S}_{cor}$ is the specific surface area corrected for the increase in weight after the modification. Determining W and $\bar{S}_{cor}$ is difficult when the inner surface of an open tubular column is modified in situ. Hence it is not possible to calculate the true coverage of bonded phases in an open tubular column and so compare it with the coverages obtained for conventional columns using phases with similar functionalities.

Using reaction (5.27) in open tubular columns it is possible to obtain alkylsiloxane, cyanoethylsiloxane, and aminopropylsiloxane stationary phases, as well as other grafted phases that cover the whole range of polarities from reversed-phase to normal-phase liquid chromatography.

By way of example we shall describe the procedure used by Tsuda et al. [75] for grafting octadecyltriethoxysilane (ODS) onto an open tubular column. The capillary is filled with xylene, and then 50 μl of 1% ODS in xylene is passed slowly through the capillary with a pressure drop of 1.8 MPa, keeping the temperature in the capillary at 95-110°C. Then nitrogen is blown through the tube while the temperature is raised from 60 to 110°C at 0.5°C/min. Finally, moist nitrogen is blown through for 6 h at 130°C and the column dried by a stream of nitrogen at 130°C for 3 h.

When the organosilane has more than one functional group, every trace of water must be eliminated from the system to prevent the silane from polymerizing. These "silanizing" procedures are clearly in need of optimization for open tubular columns. For in-

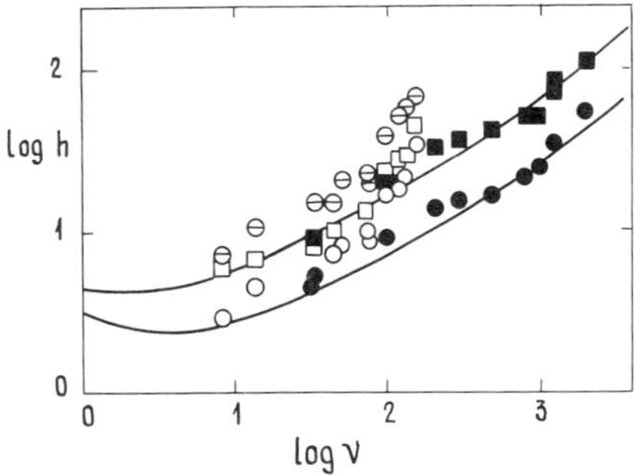

Fig. 5.10. Reduced HETP versus reduced velocity for an open tubular column without a stationary phase (k' = 0). The solid line is Taylor's equation (3.10), and the experimental points were obtained by using capillaries with diameters ranging from 50 μm to 195 μm. Data from [310].

stance, it would be reasonable to have an additional "end-capping" stage to block the residual silanol groups with trimethylchlorosilane.

A cationite coating can be obtained in an open tubular column by copolymerizing, for 20 h at 100-200°C, styrene and divinylbenzene on the surface of a quasi-silica gel porous layer and then sulfonating it, or by sulfonating a phenylsiloxane layer obtained using reaction (5.27) [81, 82, 90].

Interestingly, the poly(styrene-divinylbenzene) coating can be used for reversed-phase chromatography, too. Its advantage over alkylsiloxane phases is that it is very stable hydrolytically and can be used with eluents whose pH is greater than 8.

The grafted polyimine phase

$$-\overset{|}{\underset{|}{Si}}-(CH_2)_3-N{=}CH(CH_2)_3CH{=}\!\!\left(N(CH_2)_6-N{=}CH(CH_2)_3CH\right)_{\!n}{=}N(CH_2)_{11}CH_3$$

which is obtained as described in [85], has anionite properties.

Open tubular columns which have internal diameters greater than 60 μm and prepared as described previously have HETP's very close to those predicted by Golay [151]. This was shown by Tsuda and Novotny [310] for an unretained (k' = 0) compound (Fig. 5.10) and Tsuda et al. [75] for compounds adsorbed onto an octadecylsiloxane bonded phase (Fig. 5.11).

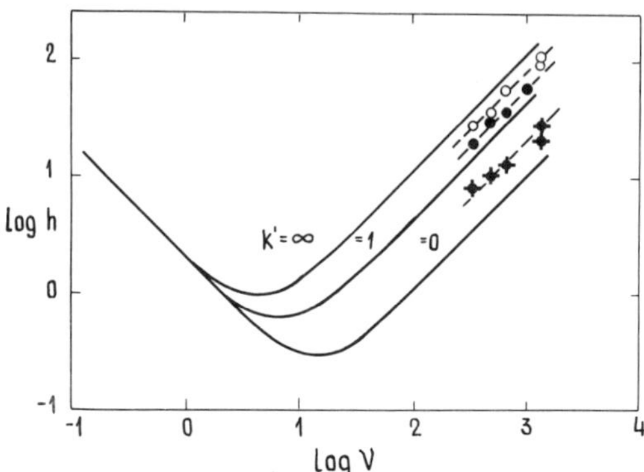

Fig. 5.11. Reduced HETP versus reduced velocity for adsorption on an octadecylsiloxane stationary phase in an open tubular column (d_c = 60 μm). The solid line was calculated from Eq. (3.9). Experimental points: o) diphenyl, k' = 1.22; ●) benzene, k' = 1.11 ✦) naphthalene, k' = 0.56. Taken from [75].

A number of important points for the estimation of the quality of open tubular columns arise from these data on the HETP versus velocity relation.

1) If for $\nu > \nu_0$ the slope of the experimental h versus ν curve is greater than the theoretical slope given by (3.9), then the stationary phase layer is too thick and is comparable to the radius of the column.

2) If the slope corresponds to the theoretical slope but the experimental points lie above the theoretical curve, then the extracolumn spreading in the system is greater than is tolerable. Remember that at large F the extracolumn system acts like an ideal mixer and so its contribution to the HETP is independent of velocity (see Section 5.1).

Since open tubular columns with diameters less than 10 μm are the most promising for liquid chromatography, we shall now look at how well their behavior corresponds to theory. Figure 5.12 shows the HETP versus reduced velocity curves obtained by Tsuda and Nakagawa [20] for 6-μm and 11-μm open tubular columns. The slope of the curve for the 11-μm column corresponds to the theoretical value, but the curve is shifted toward higher HETPs. Similar results have been obtained by Krejči [70] and Tijssen [21] for columns with internal diameters between 10 and 14 μm. Clearly, none of these investigators has managed to reduce the extracolumn spreading down to the requisite level. The experimental results for the 6-μm column were about 60

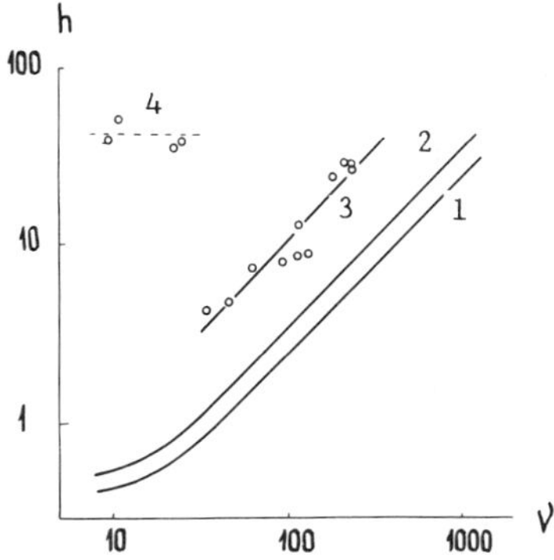

Fig. 5.12. Reduced HETP versus reduced velocity curves, for small-diameter open tubular columns. Theoretical curves: 1) k' = 0.3; 2) k' = 0.6. Experimental curves: 3) an 11-µm ID open tubular column with k' = 0.3; 4) a 6-µm ID open tubular column with k' = 0.6. (Taken from [20] with the permission of Elsevier Science Publishers.)

times higher than the theoretical values for h, but why this column had such a poor performance cannot easily be established from the data in the figure.

By using on-column potentiometric detection, Manz and Simon [103, 104, 311] were able to reduce the detection volume to a few femtoliters, and they convincingly demonstrated that Golay's theory is also valid for open tubular columns with small diameters (see Fig. 5.13).

The significant difference between open tubular and packed microcapillary columns is associated with the different ratios of stationary to mobile phases (see Table 5.1). Given the same distribution coefficients, i.e., the thermodynamic constants of sorption, the small fraction of volume taken up by the stationary phase in open tubular columns produces much smaller capacity factors k' than does the regular packing in a packed column. In order to obtain the same resolution K_R without changing the separating system, an open tubular column must have a significantly higher efficiency N than a packed column.

The expression for K_R given in Chapter 3 [(3.14)] can be written

$$N = 16\frac{K_R^2}{(\alpha - 1)^2}\left(1 + \frac{1}{\varphi k_d}\right), \tag{5.29}$$

where $\varphi = V_s/V_m$ is the volume ratio of the two phases.

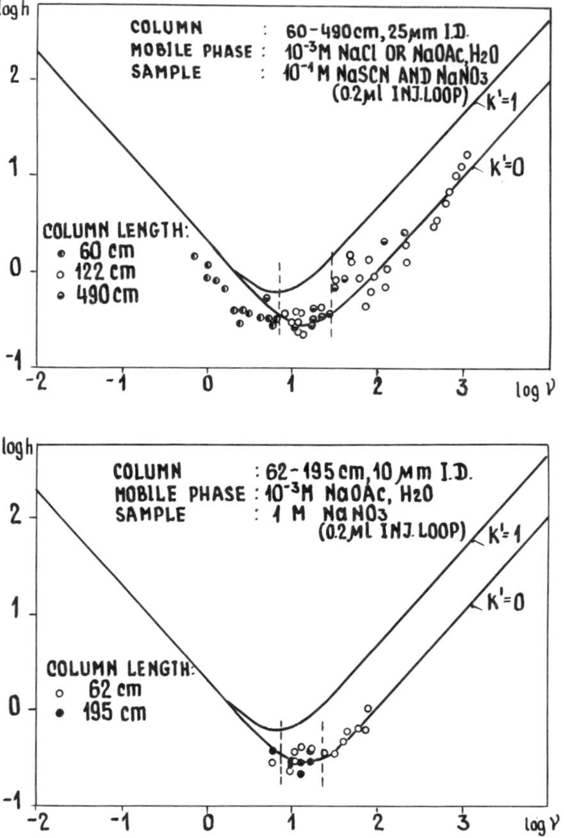

Fig. 5.13. Reduced HETP versus reduced velocity curves for open tubular colums with diameters of 25 μm (upper curve) and 10 μm (lower curve). (Reproduced from [104] with the permission of F. Vieweg & Sohn.)

The ratio between the efficiencies of a packed column and an open tubular column follows from this and depends on φ, i.e.,

$$\frac{N_{pc}}{N_{otc}} = \frac{(1+\varphi_{pc}\,k_{d_{pc}})\varphi_{otc}k_{d_{otc}}}{(1+\varphi_{otc}\,k_{d_{otc}})\varphi_{pc}k_{d_{pc}}} \,. \tag{5.30}$$

Note that a typical value of the phase ratio is 1.5-1.85 for packed columns, while that for an open tubular column is less than 0.08.

Expression (5.30) demonstrates that in order to avoid an excessive efficiency for separation the separating system must be optimized to raise K_R. Thus, if $k_{d_{otc}} = k_{d_{pc}} (\varphi_{pc}/\varphi_{otc})$, then $N_{pc}/N_{otc} = 1$.

There are two ways to raise the sorption constant: either the adsorptional activity of the stationary phase has to be increased or the eluting strength of the solvent must be reduced. Reducing the temperature also operates in the same direction for reversed-phase liquid chromatography. The best way of raising k_d is to reduce the concentration of the modifier, i.e., the strength of the solvent in the eluent. For example, Tsuda et al. [75], using an open tubular column with an ODS bonded phase, employed a significantly weaker eluent (40% acetonitrile in water) to separate polycyclic aromatic hydrocarbons than is usual for packed columns with ODS adsorbents, usually 60-70% acetonitrile being used.

It should be noted that the small phase ratio that is typical of open tubular columns also makes it impossible to utilize the steric-exclusion mechanism in them.

5.4. POSSIBLE FUTURE INCREASES IN THE PERFORMANCE OF CAPILLARY COLUMNS

As we showed in Section 5.2, the main area in which the performance of packed microbore columns can be raised is in the improvement of the packing procedures so as to reduce the thickness of the poorly packed wall region. We can expect the value of h_0 to fall to 1, which would make the efficiency of packed columns two to three times what it is now. It would seem that radially compressed packed columns based on polymeric tubes, like the radially compressed cartridges made by Waters for conventional liquid chromatography [312], are worth investigating.

For open tubular columns Giddings [95] has declared that there is potential for an increase in their selectivity and an acceleration in the mass transfer by applying external fields, i.e., by a combination of microcapillary liquid chromatography and field flow fractionation. Terabe et al. [313] have obtained some very favorable results in open tubular columns with what they call "electrokinetic chromatography." This is simply the combination of high-voltage, free-zone electrophoresis of ionic micelles and the solubilization of nonpolar molecules in the micelles. They achieved an efficiency of 250,000 theoretical plates in a 50-cm open tubular column (d_c = 50 μm) with a potential of 25 kV, which corresponds to a peak capacity of 190. Moreover, the productivity of the separating process was 148 theoretical plates per second.

On the other hand, the creation of an electroosmotic flow, given a tangential electric field, significantly diminishes the resistance to mass transfer because this flow has a flatter velocity profile than a laminar fluid flow [314, 315]. It is possible that, as Giddings has predicted, better results will be obtained with the imposition of other fields.

Attempts to overcome the drawbacks of open tubular columns associated with the low capacity and very slow flow rates by using bundles of parallel open capillaries have been made [97]. However, it is extraordinarily difficult to get experimental columns with many capillaries, each of which have the same radius.

A more promising approach is to create an entirely new sort of column, an analog of which comes from thin-layer chromatography on polyamide porous films, rather than from gas chromatography. The column we suggest is, unfortunately, just speculative at present, but may be imagined as a continuous stick of a porous stationary phase formed inside a capillary and characterized by a bimodal distribution of pore dimensions. The large pores ($d = 0.5$-$1\ \mu m$) would be analogous to the interparticle canals in conventional packed columns and through which the solvent would flow. The smaller pores ($d = 6$-10 nm) would be distributed in the thin walls of the larger pores and would provide a high sorptive capacity and rapid internal mass transfer. The system would be none other than a bundle of intersecting open tubular columns with $d_c = 0.5$-$1\ \mu m$ and $d_f = 6$-10 nm. The overall porosity could, in principle, be very high, greater than 0.9, with the porosity of the large pores ε_0 being about 0.7, and the structure still being able to maintain its integrity. Note that ε_0 in conventional packed columns is 0.35-0.45 because the packing is only sufficiently stable when densely packed.

The contribution of internal diffusional mass transfer to the HETP in this hypothetical column must be insignificant because of the thinness ($d_f = 10$ nm) of the internal diffusional path length. The largest contribution of the eddy diffusion would be $H = 2\lambda d_c = 1$-$2\ \mu m$ ($\lambda = 1$ is the coefficient of turbulent flow). Thus, the value of H_{min} should be less than 1 μm for a column of this sort, and a hydraulic permeability greater than that in packed columns.

Such a column does seem to us to be quite practicable. For example, a porous rod of silica gel could be formed directly inside a column and then hydrothermally treated to increase the pore size to $d_c = 500$-1000 nm and finally dried under mild conditions that would avoid capillary contraction, i.e., the cracking up of the silica gel. The small pores ($d_f = 10$ nm) could then be formed by impregnating the large pores with a silica sol and another desiccation.

Modern silica colloidal chemistry can accomplish all these stages. The only difficulty would be ensuring that the size distribution of the large pores is narrow. However, this difficulty

TABLE 5.7. Physical Properties of Gases, Liquids, and SC Fluids

Parameter	Gas	Liquid	SC Fluid ($T \approx T_{cr}$, $P \approx P_{cr}$)
Density ρ, g/cm^3	10^{-3}	1	$3 \cdot 10^{-1}$
Diffusivity D_m, cm^2/sec	10^{-1}	$5 \cdot 10^{-6}$	10^{-4}
Viscosity η, $g/cm \cdot sec$	10^{-4}	10^{-2}	10^{-3}

can also be overcome in principle, bearing in mind that common porous borosilicate glass is practically isoporous.

It can be said with confidence that if this sort of column can in fact be obtained, then it will only be on the capillary scale, because the isotropicity of the porous medium can only be ensured in a thin tube, the gel being formed inside one, and then aged, hydrothermally treated, and dried there.

We shall now turn to a technique that is equally promising for both open tubular and packed capillary columns: supercritical-fluid chromatography.

5.4.1. Microbore Supercritical-Fluid Chromatography

We pointed out in Sections 3.4 and 3.5 that the analysis speed and efficiency could both be significantly increased by reducing the eluent's viscosity. In fact, it is the analysis productivity N/t_R that must be increased:

$$\frac{N}{t_R} = \frac{k_0 \Delta P}{N(1 + k')} \frac{1}{h^2 \eta}. \tag{5.31}$$

Therefore, given ΔP, N, and k', and if h is kept constant, reducing η will increase N/t_R. However, it follows from

$$h\nu = \frac{k_0 \Delta P d^2}{N(\eta D_m)} \tag{3.29}$$

that in order to keep $h\nu$ constant (or only slightly changed), the product ηD_m must also remain constant (or be slightly changed). This is what happens when supercritical-fluid chromatography is used in place of liquid chromatography.

TABLE 5.8. Properties of Compounds Suitable for SC-Fluid Chromatography

Compound	bp, °C	T_{cr}, °C	P_{cr}, kg/cm^2	ρ_{cr}, g/cm^3
H_2S	−60.8	100.4	88.9	0.349
SO_2	−10.1	157.5	77.8	0.524
SF_6	–	45.6	37.1	0.750
CO_2	−78.5	31.3	72.9	0.448
NH_3	−33.4	132.3	111.3	0.235
H_2O	100.0	374.4	226.8	0.327
C_5H_{12}	36.3	196.6	33.3	0.232
C_6H_{14}	69.0	234.2	29.6	0.234
CF_3Cl	−81.4	28.8	39.0	0.58
CF_2Cl_2	−29.8	111.7	39.4	0.55
$CHFCl_2$	8.9	178.5	51.0	0.522
CCl_3F	23.7	196.6	41.7	0.554
CH_3OH	64.7	240.5	78.9	0.272
C_2H_5OH	78.4	243.4	63.0	0.276
iso-C_3H_7OH	82.5	235.3	47.0	0.279
$C_2H_5OC_2H_5$	35.6	193.8	35.5	0.264
$C_2H_5NH_2$	16.6	183.2	55.5	–
$(CH_3)_2CO$	56.2	235.5	46.6	0.273

A supercritical (SC) fluid is the condensed state above the critical temperature and pressure, i.e., the operating pressure is some two to four times the critical pressure and the operating temperature 1.2 to 1.5 times the critical one. The physical properties of gases, liquids, and SC fluids are given in Table 5.7.

The solubility of a solute in an SC fluid is important. Giddings [316] has shown that if the dissolving ability of a liquid can be estimated using Hildebrand's parameter δ_L

$$\bar{\delta}_L = (\Delta E/V_L)^{1/2}, \tag{5.32}$$

where ΔE is the molar enthalpy of evaporation, and V_L is the molar volume of the liquid, then the solubility parameter for an SC fluid δ_F is a fraction of δ_L and is given by the ratio of the densities of the SC fluid and liquid, i.e.,

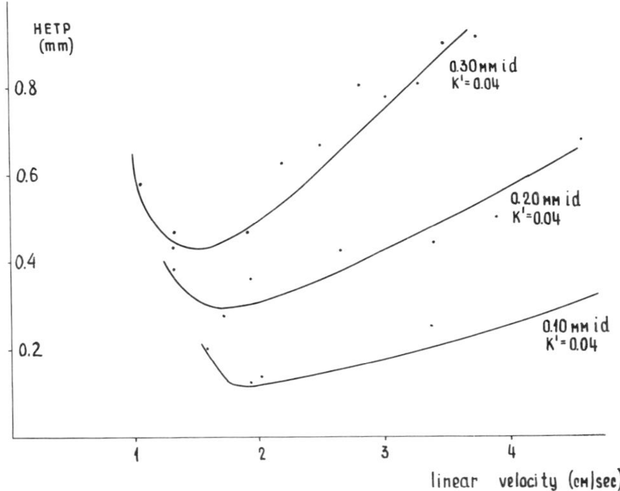

Fig. 5.14. HETP versus elution velocity for SC-fluid chromatography on open tubular columns. (Reproduced from [139] with the permission of the American Chemical Society.)

$$\bar{\delta}_F = \bar{\delta}_L(\rho_F/\rho_L). \qquad (5.33)$$

Thus, the solubility parameter of an SC fluid is heavily dependent on its density, and for large densities, created by large pressures, may approach that of a liquid.

According to Hildebrand's theory, dissolution occurs if the difference between δ for the solvent and the solute lies between +1 and −1. Thus, by varying the pressure of an SC fluid its dissolving power can be significantly changed. The properties of several compounds that can be used for SC-fluid chromatography are given in Table 5.8 [318].

Since Klesper [317] suggested it, supercritical-fluid chromatography has been developed using the apparatus of gas chromatography (eluent delivery method, sample injection). However, depending on what the decompressed state of the SC fluid was, i.e., gas or liquid, a liquid or gas chromtography detector was used. Comparatively recently, the technique began to be developed using the apparatus of liquid chromatography (miniature pumps for eluent delivery and miniature injectors for the samples). A microbore column can be easily kept at the required temperature by placing it in a thermostat developed for gas chromatographic columns. Hence, using microbore columns that have good heat conductivities simplifies the apparatus for SC-fluid chromatography and makes it as accessible as liquid chromatography.

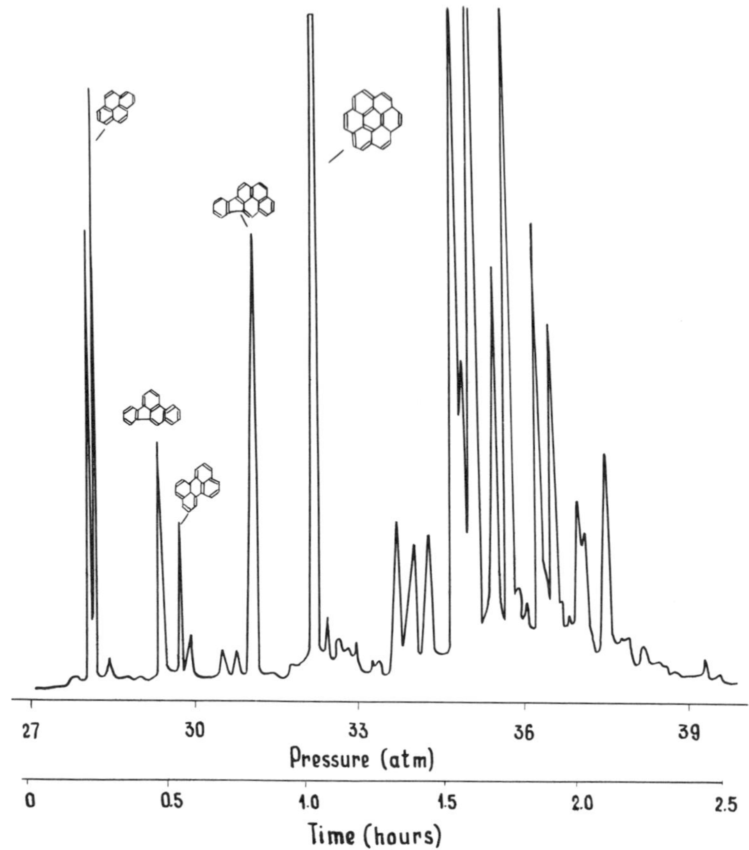

Fig. 5.15. Separation of carbon black on an open tubular column using SC-fluid chromatography. (Reproduced from [139] with the permission of the American Chemical Society.)

Readen et al. [139] have developed the technique in open tubular columns, while Hirata and Nakata [140] and Takeuchi et al. [141] have used ultramicrobore packed fused-silica columns. The open tubular technique has a smaller pressure drop down the column and less of the band spreading due to uneven velocity profiles than packed columns suffer from.

Readen et al. used a setup in which the eluent was delivered by the Varian 8500 syringe pump, which was modified for controlling the pressure as described in [139]. The column was placed in the Hewlett—Packard 5700A gas chromatograph's thermostated chamber. The sample was introduced either with a Carlo Erba on-column injector or with a Valco rotary valve, which have volumes of 0.5 and 0.2 µl, respectively. In order to obtain the necessary mobile phase temperature the eluent was passed through a 1000 × 0.76 mm stainless steel capillary inside the thermostat before entering the column.

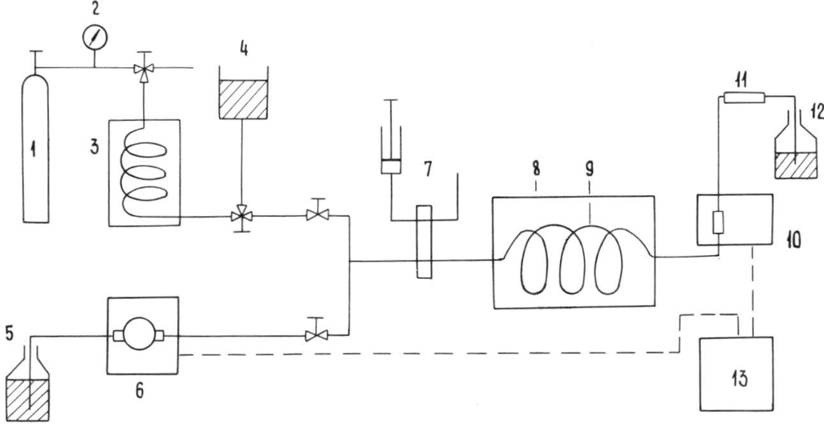

Fig. 5.16. Diagram of an SC-fluid chromatograph on packed ultramicrobore columns. 1) Cylinder with nitrogen; 2) manometer; 3) pump A; 4, 5) solvent reservoirs; 6) pump B; 7) microloop injector; 8) oven; 9) column; 10) UV detector; 11) pressure throttle; 12) overflow; 13) recorder. (Reproduced from [140] with the permission of Elsevier Science Publishers.)

A Valco valve, which has a 0.2-μl sampling loop, was used to split the flow into 1:1 and 1:2 ratios. The necessary pressures were maintained in the column by a back pressure of compressed nitrogen. The separation column itself was an open tubular Pyrex capillary with an internal diameter of 0.1-0.3 mm, coated with poly(methylphenylsiloxane).

The detector was a Perkin–Elmer 240A spectrophotometer, the flow cell simply being pieces of silica capillary 1 cm long, and either 0.3, 0.2, or 0.1 mm internal diameter with the corresponding volumes of 0.71, 0.31, and 0.08 μl. The exciting wavelength was 280 nm. The working load on the column was 10-100 ng per component. The eluent was n-pentane with pressure programming up to 4 MPa and a temperature of 210°C.

Figure 5.14 demonstrates the relationship between the HETP and the elution velocity for naphthalene. It can be seen that, as predicted by theory, the HETP falls when an open capillary with a smaller diameter is used. According to Golay's equation (3.9), the HETP minimum should occur for $0.289d_c$, but the experimental value is 3-3.5 times this value, which may be due to the considerable spreading in the extracolumn system (the detector, splitter, and injector).

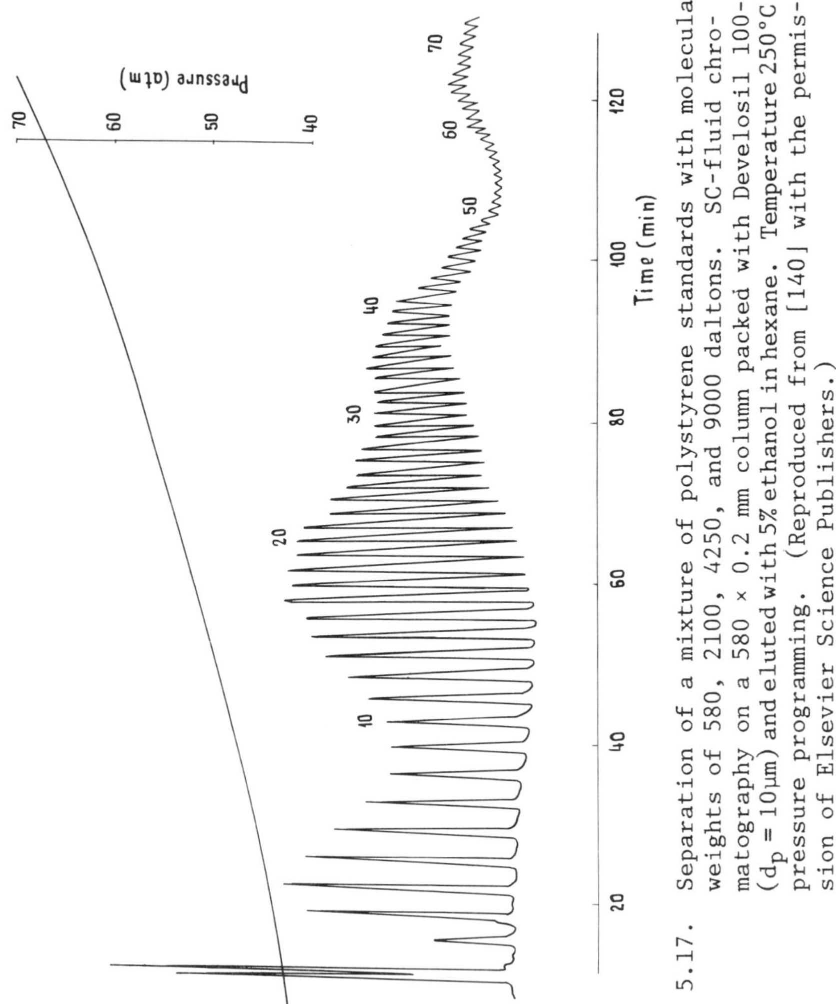

Fig. 5.17. Separation of a mixture of polystyrene standards with molecular weights of 580, 2100, 4250, and 9000 daltons. SC-fluid chromatography on a 580 × 0.2 mm column packed with Develosil 100-10 ($d_p = 10\mu m$) and eluted with 5% ethanol in hexane. Temperature 250°C and pressure programming. (Reproduced from [140] with the permission of Elsevier Science Publishers.)

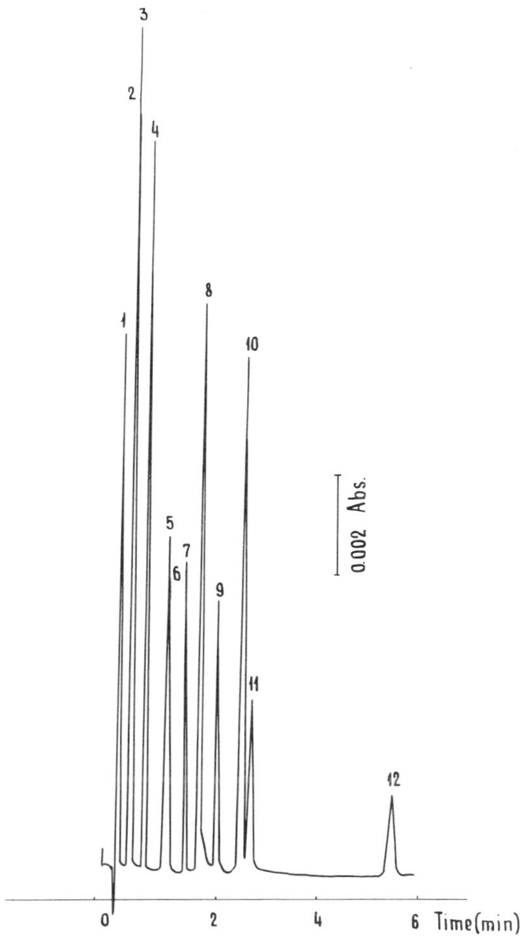

Fig. 5.18. Separation of polycyclic aromatic hydrocarbons by SC-fluid chromatography with carbon dioxide as the mobile phase. A 150 × 0.34 mm column packed with ODS-SC-01 (d_p = 5 μm) was used, temperature 35°C, pressure 15 MPa. (Reproduced from [141] with the permission of Elsevier Science Publishers.)

For a linear velocity u of 4 cm/sec (about 10 times the optimal value) the open tubular column had an efficiency of $\sim 10^4$ theoretical plates and a productivity N/t_R of about 40 plates/sec.

The system was used to analyze extracts of carbon black (Fig. 5.15). The analysis time of 120 min was somewhat longer than that when analyzing the products using high-performance capillary liquid chromatography. However, the results of Readen et al. can hardly be used to judge the possibilities of open tubular column SC-fluid chromatography because the experimental conditions were far from optimal.

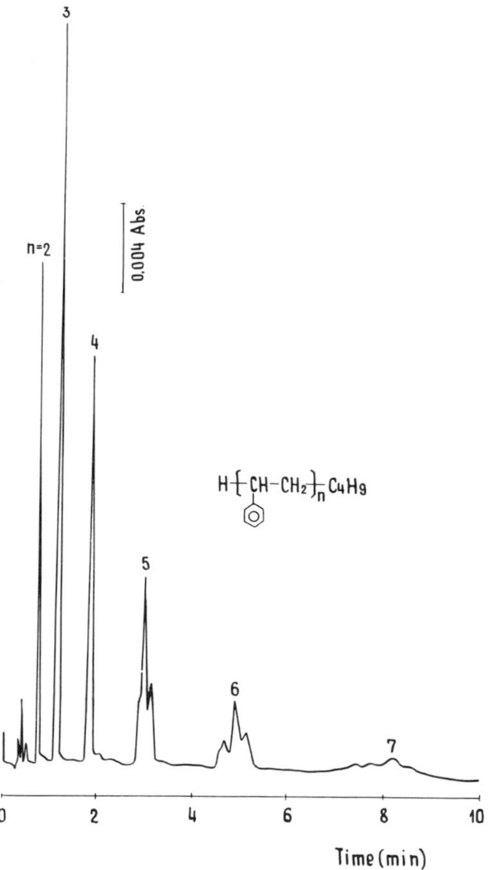

Fig. 5.19. Separation of styrene oligomers using SC-fluid chromatography on a capillary column. Same conditions as in Fig. 5.18. UV detection at 205 nm. The sample was a polystyrene standard with $\bar{M}_W = 500$. (Reproduced from [141] with the permission of Elsevier Science Publishers.)

Hirata and Nakata [140] used the equipment shown in Fig. 5.16. The separating column was a 580 × 0.2 mm fused-silica capillary filled with Develosil 100-10 and was placed in a Shimadzu GC-5A gas chromatography thermostat. The detector was an absorbance one (JASCO's UVIDEC 100 III). The eluent was delivered to the column by two pumps: A, which provided a constant solvent flow, and B, which was flow programmed.

This system has been used for analyzing polycyclic aromatic hydrocarbons and oligostyrenes (Fig. 5.17). Hirata and Nakata demonstrated that the HETP for dibenz[a,c]anthracene in a 1:9 mixture of ethanol and hexane is practically independent of the elution velocity, for velocities between 1.5 and 1.8 mm/sec. Below the critical temperature all the polycyclic hydrocarbons were eluted together, which can be thought of as adsorption chromatography with an eluent that has a high elution strength. As the temperature was raised above the critical point, the retention volume first increased rapidly and then slowly fell.

Hirata and Nakata used the oligostyrenes to study the effect of adding modifiers, such as ethanol, propanol, dioxane, diisopropyl ether, ethyl acetate, and tetrahydrofuran, to the basic mobile phase of hexane. Their results provide evidence that the basic separation mechanism of the oligomers depends on the changes in solubility in the eluent caused by changes in pressure. They were also able to set new records for the resolution (70 oligomer homologs were identified) and speed of the separation, records both for supercritical chromatography and for all other oligomer separation techniques.

The system Takeuchi et al. [141] have developed used supercritical carbon dioxide. It was supplied from gas cylinders and delivered by a Familic 300S pump (JASCO). The pump head was cooled by dry ice. The UVIDEC 100-II (JASCO) UV detector was used, with the cell kept under pressure. The column, with an internal diameter of 0.34 mm, was made from fused silica and was glued into a steel capillary with epoxy resin. The chromatogram in Fig. 5.18 shows the separation of 12 polycyclic aromatic hydrocarbons on ODS. A comparison of these results, obtained using SC-fluid chromatography, and those obtained using conventional reversed-phase chromatography indicates that the retention of polycyclic aromatic hydrocarbons is not only determined by the dissolving power of the eluent (supercritical carbon dioxide), but also depends on how the solute interacts with the sorbent. This is supported by the SC-fluid chromatography of styrene oligomers, which separated the oligomer homologs according to position isomers, which is typical of the adsorption chromatography of oligostyrenes (Fig. 5.19).

Chapter 6

APPLICATION OF MICROBORE CHROMATOGRAPHY TO THE ANALYSIS OF ORGANIC AND INORGANIC COMPOUNDS

6.1. AMINO ACIDS, THEIR DERIVATIVES, PEPTIDES, AND PROTEINS

6.1.1. Ultrasensitive Analyses of Amino Acids and Their Derivatives with Fluorometric Detection

Raising the analysis sensitivity to amino acids and their dansyl derivatives is important for determining the primary structures of proteins, i.e., their amino-acid composition and sequence. A modern amino-acid analyzer, consisting of an ion-exchange system, a postcolumn derivatization unit for reacting the amino acids with o-phthalic dialdehyde, and a fluorometer, has a detection limit of 10^{-12} mole. This level is also the lowest possible for dansylated amino acids using thin-layer or high-performance chromatography on silica gel or alkylated silica gel.

Microbore liquid chromatography can raise this sensitivity by more than two orders of magnitude [12, 133]. Dansylated amino acids can be separated in a PTFE or glass column 250 mm long and inside diameter 0.5 mm containing a filter of porous titanium or polyamide. The column is packed with octadecylsilylated silica gel Silasorb 300-C_{18} (sold by Lachema, Czechoslovakia). A dual gradient was set up: a pH gradient and a concentration gradient of acetonitrile. The acetonitrile is delivered to the column by two automatically controlled syringe pumps, which provide a total delivery of 1 ml/h. In order to program the eluent reproducibly it was found essential to combine the solvents from the two pumps in a dynamic mixer about 10 μl in volume and agitated with a miniature magnetic armature. The effluent was detected in a fluorometer with a cell 1.3 μl in

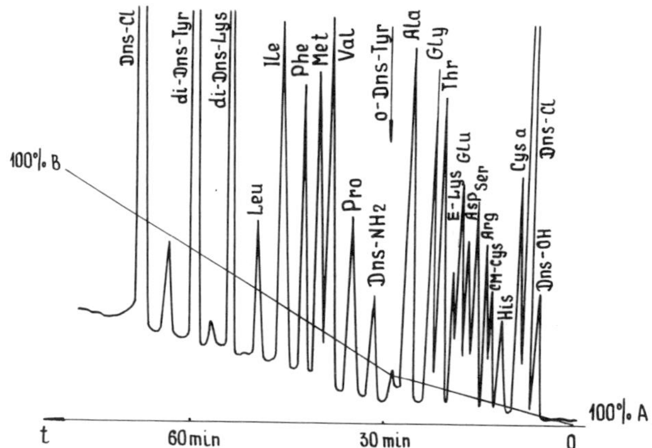

Fig. 6.1. Microbore liquid chromatogram of DNS amino acids taken on a 250 × 0.5 mm column packed with Silasorb 300-C_{18} (d_p = 7 μm). Eluent A was 25% acetonitrile and 75% 0.01 M sodium formate, pH 3.5. Eluent B was 60% acetonitrile and 40% 0.01 M phosphate buffer, pH 7.0. Elution gradient was from 100%A to 100%B, velocity 1 ml/h, $2\text{-}5 \cdot 10^{-13}$ mole of each DNS amino acid was introduced, UV fluorescence detection (excitation at 380 nm, and detection at 570 nm). (Reproduced from [133] with the permission of Elsevier Science Publishers.)

volume, the exciting radiation being the UV output from a DNU-65 deuterium lamp fitted with a UV-1 separating filter and a ZhS-16 blocking filter.

This equipment yields an analytical sensitivity of around 10^{-13} mole. The DNS amino-acid peaks take up 40-80% of the recorder scale with a noise level of about 0.5%. Thus, the threshold detection (a signal double the noise level) is $(1\text{-}3) \cdot 10^{-15}$ mole. Figure 6.1 is a chromatogram of a mixture of DNS amino acids that was obtained by gradient elution with an analysis time of 1 h.

Mal'tsev et al. [12, 133] have developed an analysis method that combines high sensitivity with excellent reproducibility, achieving relative standard deviations of 1.7% with respect to retention time and 2.5% with respect to normalized peak height. It has been shown that the application of rapid precolumn derivatization with dansyl chloride (dansylation) [319] makes it possible to use microbore techniques for amino-acid analyses and get the same level of sensitivity. Since it is, in practice, difficult to dansylate a few microliters of sample and then introduce it into a microbore column without some quantitative loss, the sample was concentrated directly in the analytical column. It was established that the column efficiency (N) and resolution (K_R) were only slightly depressed when samples in volumes up to 100 μl were introduced in a 10% solution of acetonitrile in 0.1 N hydrochloric acid. For instance, K_R fell from 0.98 to 0.94 for aspartic and glutamic acids, while N fell from 7300 to 6600 for alanine. In fact, all DNS amino acids remain practically uncharged for pH values less than 1, and hence they are concentrated in the initial sections of a microbore column. Thus, the volume of the sample can be increased to 100 μl without worsening the separation.

Takeuchi et al. employed fused-silica ultramicrobore columns to separate DNS amino acids [320]. They employed reversed-phase chromatography with either isocratic modes (with a column 250 × 0.34 mm and ODS Hypersil sorbent) or gradient elution (with a column 100 × 0.34 mm and the same sorbent). The mobile phases used were mixtures of acetonitrile and 0.13 M ammonium acetate at flow rates of 8 and 4.2 μl/min for isocratic and gradient regimes, respectively. The detector was a spectrophotometer tuned to 222 nm, and the samples were fed to the column in amounts of 20 pmoles. Takeuchi et al. obtained threshold sample masses of 70-310 fmoles, which is significantly greater than those reported in [133]. The analysis took 25 min. Only 18 of the 25 amino acids found in protein hydrolyzates could be identified. Takeuchi et al. demonstrated, however, that the separation of weakly retained components could be improved with gradient elution. They used the method to determine the amino-acid composition of soy sauce and sake, and showed that 60 pmoles of amino acid could be detected in 11 nl of sake.

6.1.2. Microbore Chromatography of PTH Amino-Acid Derivatives

The amino-acid sequences of peptides and proteins can be determined by identifying the PTH derivatives of the amino acid produced by Edman's degradation.

Takeuchi et al. [321] employed isocratic reversed-phase chromatography in fused-silica capillaries 404 × 0.34 mm with ODS SC-01 silica gel (d_p = 5 μm). They used a 28.8:6:64.6 mixture of acetonitrile, tetrahydrofuran, and 0.005 M sodium acetate (pH 5.15) with an eluent flow rate of 13 μl/min. The sample was 0.02 μl of a 0.003-0.004 M solution of PTH amino acids and was detected by a photometer at λ = 254 nm. This method can separate 21 amino acids in 20 min, but it cannot distinguish between the members of the following pairs: tryptophan and phenylalanine, methionine and valine, glycine and histidine, or cysteine and aspartic acid.

Cunico et al. [322] have developed another sensitive method for analyzing PTH amino acids in 150 × 1 mm columns with a nitrylalkylsiloxane bonded phase. The elution was achieved with a preformed gradient at a flow rate of 50 μl/min. Twenty PTH amino acids were separated in 40 min. The threshold sample mass was 0.5-1 pmole, and so this method is 1-1.5 orders of magnitude more sensitive than those using the usual 150 × 4.6 mm microbore columns.

6.1.3. Ultrasensitive Analyses of Amino Acids and Peptides with a Postcolumn Ninhydrin Reaction

This form of analysis is a modification for capillary columns of the method devised for analyzing amino acids by Spackman, Stein, and Moore [323], which is the one embodied in most commercial automatic amino-acid analyzers. A glass 150 × 0.5 mm microbore column packed with the cationite Aminex A-7, Bio-Rad (d_p = 8 ± 2 μm), or Durrum DC-5A (d_p = 6 ± 1 μm) is at the heart of the system. Note that the columns in conventional analyzers are at least twice this length. The flow reactor is made from a section of PTFE capillary 5 cm long and 0.2 mm inside diameter and is heated to 160°C. The column and ninhydrin feed line are connected to the reactor via a T-junction mixer 1 μl in volume. The ninhydrin is fed in at 63 μl × h^{-1} for an amino-acid analysis and at 125 μl/h for a peptide analysis. A preformed gradient elution is used with step increments in the pH for an amino-acid analysis, while both the pH and the buffer's ionic strength are increased for the peptide analysis. The elution rate is 125 μl/h.

By optimizing the composition of the ninhydrin reagent [222] and carrying out the reaction at a high temperature (160°C) the reaction time can be shortened to 1 min, and hence the length of the reactor, which determines the extracolumn dispersion, can be reduced to a minimum. The colored products of the reaction are detected with a spectrophotometer, whose cell volume is 1 μl, set for a wavelength of 570 nm (except for proline, where the wavelength is 440 nm). To keep the solution from boiling in the reactor, a back-pressure restrictor, which makes the pressure inside the reactor rise to 0.3 MPa, is placed at the exit of the reactor tube.

A chromatogram of an artificial mixture of $5 \cdot 10^{-11}$ mole of each of the acidic and neutral amino acids is given in Fig. 6.2. Since the peak height is 50% of the scale of the recorder and the noise level is about 1% of the scale, the threshold detection is $2 \cdot 10^{-12}$ mole (for a signal twice the noise level). Note that the short-period noise is much lower in amplitude, i.e., about 0.1%. The long-period noise fluctuations are related to the inefficient mixing of the ninhydrin and effluent due to the small flow rates of microbore chromatography. Packing the reactor with glass microspheres about 10 μm in diameter significantly reduces the long-period noise.

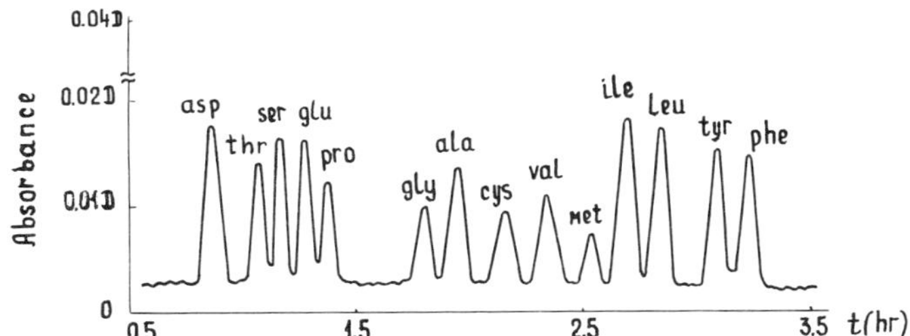

Fig. 6.2. Microbore liquid chromatogram of a standard mixture of acidic and neutral amino acids taken on a 150 × 0.5 mm column packed with Aminex A-7 (d_p = 8 ± 2 μm). Eluent: a citrate buffer with 1) 0.2 N Na^+, pH 2.8, 5% n-propanol (90 μl); 2) 0.2 N Na^+, pH 3.25 (160 μl); 3) 0.2 N Na^+, pH 4.25, 9% n-propanol (125 μl). Flow rate 125 μl× h^{-1}, temperature 50°C. A 5-μl sample containing $5 \cdot 10^{-11}$ mole of each amino acid was injected. Fluorometric detection at 570 nm after ninhydrin reaction [223].

A chromatogram of a supersensitive analysis of some peptides is given in Fig. 6.3. Under the conditions described above, 13 peptides were separated from the products of the trypsin hydrolysis of oxidized ribonuclease A (1 μg of the protein). The number of peaks registered corresponded to the theoretical number of peptides that should result from the hydrolysis; trypsin is specific for cleaving the peptide links adjacent to arginine and lysine residues. The sensitivity of the method can be substantially increased by using o-phthalic dialdehyde and a fluorometer (V_d = 1.3 μl) instead of ninhydrin in the postcolumn reactor. The sensitivity attained with this setup for the amino-acid mixture was 10^{-13} mole and for the trypsin hydrolysis 0.1 μg of protein. These results demonstrate that optimizing for the analysis sensitivity (shortening L and reducing u to u_{opt}) can yield sizable gains in sensitivity even when postcolumn derivatization is used, i.e., when a significant amount of extracolumn spreading is introduced. However, the price of the sensitivity gain is a longer analysis time. Thus, a modern automatic amino-acid analyzer takes 30-40 min to analyze the products of a protein hydrolysis, whereas the microbore apparatus will take 3.5 h. However, if the primary structure of a rare protein is to be studied, then a longer analysis time will be admissible for the sake of reducing the sample mass.

6.1.4. Ultrasensitive Analyses of Amino Acids and Amines (Postcolumn Derivatization with o-Phthalic Dialdehyde)

Kucera and Umagat [172] have reported an ultrasensitive microbore analysis of amino acids and amines with a postcolumn reaction

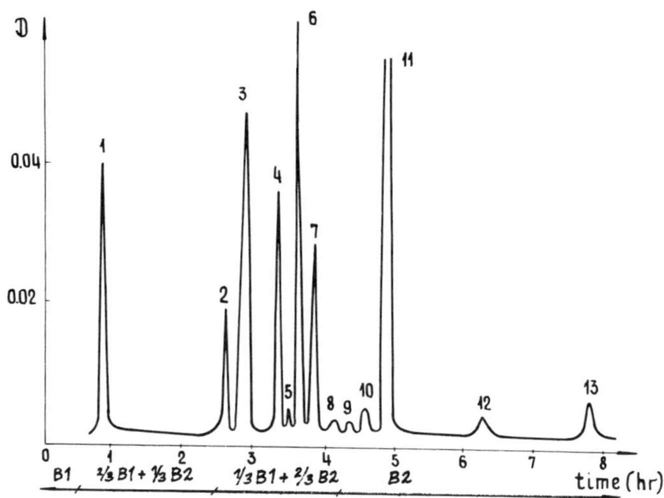

Fig. 6.3. Microbore liquid chromatogram of a trypsin hydrolyzate of 1 μg of oxidized ribonuclease taken on a 150 × 0.5 mm column packed with Aminex A-7 (d_p = 8 ± 2 μm). Eluent: 1) 70 μl of a citrate buffer, pH 3.0, 0.2 N Na^+ (buffer B1); 2) 250 μl of 2/3 B1 + 1/3 B2; 3) 250 μl of 1/3 B1 + 2/3 B2; 4) 1400 μl of an acetate buffer, pH 5.0, 2.0 N Na^+ (buffer B2). Other parameters as in Fig. 6.2 [223].

with o-phthalic dialdehyde and a fluorescence detector. The column was 500 × 1 mm and filled with Alltech ODS sorbent (d_p = 7 μm). Five amino acids were separated in 32 min in a 98:2:0.2 mixture of 0.003 M KH_2PO_4 (pH 3), acetonitrile, and tetrahydrofuran. A 0.2-μl sample with a concentration of $2 \cdot 10^{-7}$ mole/ml was injected and a sensitivity limit of about 10^{-15} mole obtained. The postcolumn reaction took 6 sec in a special reactor, 30 nl in volume.

6.1.5. Microbore Steric-Exclusion Chromatography of Large Peptides

One of the most important and as yet most difficult steps in the determination of the primary structure of a protein is the separation of the mixtures of large peptides that are obtained by cleaving the protein. The accepted methods for doing this are steric-exclusion and reversed-phase chromatographies [324]. Such advantages of capillary chromatography as high sensitivity, rapid analysis, and economy, which enable the conditions for a preparative separation to be selected rapidly, make the technology very attractive when dealing with this class of compounds.

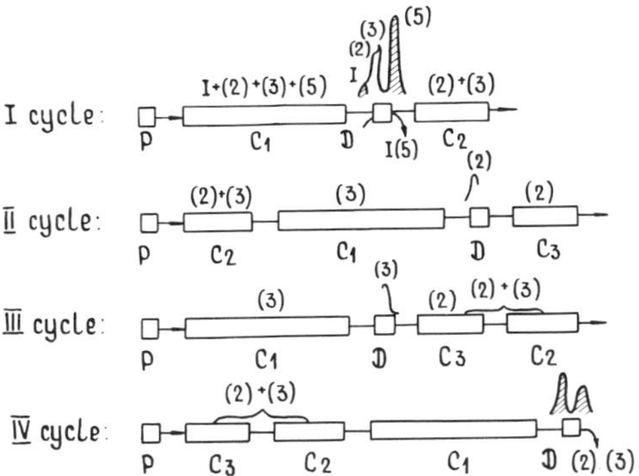

Fig. 6.4. Recirculation steric-exclusion microbore liquid chromatography of a mixture of insulin (1), its A-chain (2), and B-chain (3), and the low-molecular-weight products (5). P) Pump; D) detector; C) column. The curves shown above the detector are the chromatogram peaks in that cycle, the shaded peaks being discarded after that cycle [325].

The separation of peptides with very similar molecular masses requires a column with a very high selectivity. The most important way of raising the selectivity is to use cascades or the multidimensional technique.

A system of rotary valves is used to govern the movement of substance around the cascade network of separating columns, when they are the conventional diameter. This approach is intolerable for capillary columns because of the significant extracolumn spreading that takes place in the valves and connecting capillaries.

Gankina et al. [325-327] have described cascaded columns made from polyethylene capillaries. These can be easily connected together and to the inlet and outlet of the detector. Thus, the spreading due to the connections can be practically eliminated.

If the components are to be separated further, in one or more columns in series, after the chromatography, then they are transferred to a column connected to the exit of the detector. This column is then disconnected from the detector and either connected to a new system of columns or eluted with a new solvent. Either a single microbore chromatograph can be used, with the bottom column reconnected to the top each time, or several column–detector units (the working modules) can be used at the same time, as is usual in cascade chromatography.

Two mixtures, the products of denatured, reduced, and carboxymethylated insulin (a mixture of insulin's A and B chains) [325, 327] and the products from the partial cleaving of cytochrome C by cyanogen bromide, as described in [328], were used to test the method. The latter mixture contains five peptides which were eluted in the following order (given in terms of their relative molecular mass and the number of amino-acid residues): 1) 8000, 1-80; 2) 6500, 1-65; 3) 3800, 66-104; 4) 2300, 81-104; 5) 1400, 66-80. The peptides were completely separated with a high-performance 250 × 5 cm column. The packing was Sephadex G-50 and the eluent a 10% aqueous solution of formic acid. The separation took 72 h [328].

The chains of the reduced insulin have also been separated using microbore steric-exclusion chromatography [325]. This was carried out on three polyethylene columns, each 0.85 mm in diameter and 28, 12, and 11 cm long, respectively, packed with Sephadex G-50 and eluted with 5 M Gu·HCl. The separation strategy is shown in Fig. 6.4. Column C_1 is fixed and connected to the detector, while columns C_2 and C_3 are replaceable. Column C_2 was connected to the detector outlet in order to collect the separated components, and column C_3 was connected in front of C_1 directly to the injector. After the required fraction had been transferred, under the control of the detector, to C_2, it was disconnected and reconnected to the injector in front of C_3. Thus, a separation cycle was implemented on the three columns. If the volume of the recycle fraction was too large, then both columns C_2 and C_3 were connected in series after the detector. These columns were then connected in front of the fixed column for the next separation cycle. The complete separation required four cycles (4 h). Gankina et al. [325] have shown that the separation of this mixture using conventional recycle chromatography required 40 h.

However, the use of 5 M Gu·HCl as the eluent means the fraction has to be desalted before it can be used. The desalting is done by passing the requisite fraction through the detector and into a 215 × 1 mm column packed with Sephadex G-10 which has been neutralized with ammonia solution (pH 9). The column is then eluted with the same buffer solution at 1 μl/h under the control of a detector, and after the first peptide peak has been collected in an ampule it is lyophilically dried for further investigation.

Gankina et al. [326, 327] have also separated the fragments of the partial cyanogen bromide digestion of cytochrome C using four 0.85-mm ID columns, two (C_1 and C_2) packed with Sephadex G-50 and each 22 cm long, and two (C_3 and C_4) packed with Spheron P-1000 and each 11 cm long. Columns C_1 and C_2 separated the mixture into three fractions with a 10% aqueous solution of formic acid, i.e., peptides 1 and 2 poorly separated, peptides 3 and 4 separated together, and peptide 5 separated alone. Peptide 5 was well separated and so taken off

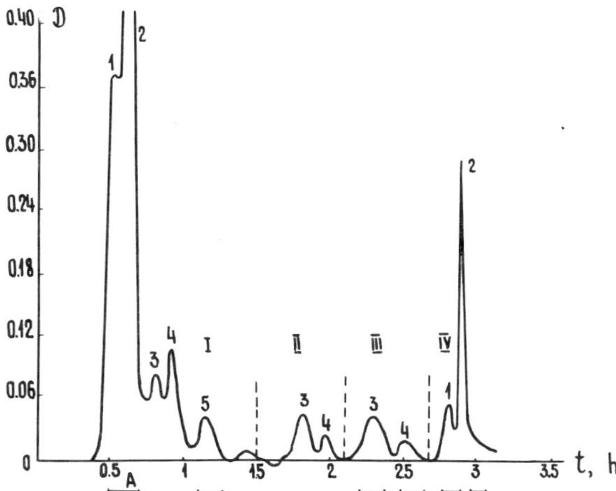

Fig. 6.5. Cascade microbore liquid chromatography of a partial cyanogen bromide digestion of cytochrome C using two 220 × 0.085 mm columns packed with Sephadex G-50 (d_p = 20-30 μm) and two 110 × 0.085 mm columns filled with Spheron P-1000. Eluent: 10% aqueous formic acid for the Sephadex and 10% aqueous formic acid with 3% tert-butanol for the Spheron packing; the flow rates were 250 μl/h and 2 ml/h, respectively. 20 nmoles of hydrolyzate were injected. The dashed lines separate the different cycles. Cycles I-III were on Sephadex, and cycle IV on Spheron. The lines under the peaks indicate the separated peaks [327].

straight away. In order to separate completely peptides 3 and 4, C_2 was disconnected from C_1 and reconnected to the outlet of the detector. The peptides were then separated using the recycle technique and were taken out after the third cycle. The poorly separated 1 and 2 peptides were separated in columns C_3 and C_4 (containing Spheron P-1000) with a 10% aqueous solution of formic acid to which 3% tert-butyl alcohol had been added. To begin with, C_4 was connected to the detector and washed with the eluent in order to stabilize the base line. Column C_3 was then connected in front of C_4, and in this way peptides 1 and 2, which were in C_3, were reintroduced into the system. The peptides contained in C_3 were fully separated after only one cycle with an elution rate of 250 μl/h. The profile of the complete separation is shown in Fig. 6.5, and the whole procedure took only about 3 h, i.e., 18 times faster than the separation on conventional columns [328].

This separation required two packings (Sephadex and Spheron) and the recycling technique. In order to improve the separation of peptides that have similar molecular weights and which are poorly separated by steric-exclusion chromatography, it is possible to use other techniques such as adsorption or reversed-phase chromatography.

When peptides are detected by their absorbance at 280 nm, the sensitivity is determined by their content of aromatic amino-acid residues. Hence, while an analysis of the insulin fragments required a sample of 1-1.5 nmoles (the peak heights after the first cycle taking up to 80% of the recorder scale, and 20-30% after the fourth cycle), more than 10 nmoles was needed for the cytochrome hydrolyzate analysis because of the different number of aromatic amino-acid residues in the peptides (the peak heights ranged from 2% to 80% of the scale).

A cascade steric-exclusion microbore analysis with three or four cycles requires between 1 and 10 nmoles of peptide. However, it should be noted that a sixfold increase in the quantity of peptide did not significantly worsen the separation.

6.1.6. Reversed-Phase Chromatography of Proteins in Short Microbore Columns

Nice et al. [329] have investigated whether short microbore columns can be applied to proteins. The basic problem of high-performance liquid chromatography for separating or purifying proteins is related to the low concentrations of the protein in the effluent. These concentrations are usually only a few μg/ml, and hence it is difficult to process the fractions with the aim of determining the protein's amino-acid sequence or for calculating the specific activity of radioactively labeled proteins. Improving the performance of the columns for the reversed-phase chromatography of proteins, and thus increasing the concentration of the proteins in the effluent, by slowing down the elution rate also lowers the yield of hydrophobic proteins [330]. The other way of concentrating effluents for this sort of chromatography is to use volatile eluents. However, effluent evaporation can chemically modify some proteins, even if it occurs in an inert atmosphere. Moreover, the yield can also be reduced due to nonspecific adsorption of the protein on the vessel walls when the organic solvent is removed [331]. Since a peak's volume is proportional to the length and to the square of the diameter of the column, the effluent from a short (10 cm), high-performance microbore column will be much more concentrated than that from a conventional column (250 × 4.6 mm), given the same linear mobile-phase velocity. By using the large capacity of a reversed-phase or ion-exchange column a large volume of protein solution can be concentrated directly in the analytical microbore column using gradient elution which, by the way, gets around the problem of small input samples.

Nice et al. [329] have applied microbore columns filled with Hypersil ODS (5 μm) and 2.1 or 1 mm in diameter and 7.5 cm long ($N > 30{,}000$ plates) to the analysis or concentration of proteins that are only available in microgram or nanogram quantities. The

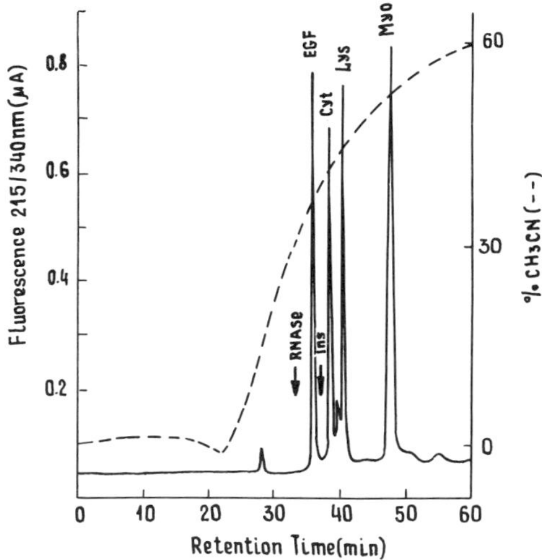

Fig. 6.6. Microbore liquid chromatogram of a mixture of standard proteins on a 75 × 2.1 mm column filled with Hypersil ODS (d_p = 5 μm). Eluent: A) 0.9% NaCl, pH 2.1; B) 60:40 acetonitrile and 0.9% NaCl, pH 2.1. The gradient profile is shown in the figure as a dashed line. Elution flow rate was 100 μl/min; 2 ml of sample was injected, and it contained 180 ng of EGF (epidermal growth factor), 450 ng Cyt (cytochrome C), 550 ng Lys (lysozyme), and 750 ng Myo (myoglobin). The positions of RNase (ribonuclease) and Ins (insulin) are indicated by arrows. (Reproduced from [329] with the permission of Elsevier Science Publishers.)

method yields proteins concentrated enough to determine their amino-acid sequences.

The concentration method also makes it possible to determine the specific activity of radioactively labeled proteins. Figure 6.6 shows the reversed-phase chromatogram of a mixture of standard proteins. The solute was detected by measuring the fluorescence of tryptophan excited by a wavelength of 215 nm and measured at 340 nm. The threshold sample mass came to 5 ng, though for UV detection this figure can be reduced to 1 ng.

6.2. STERIC-EXCLUSION MICROBORE CHROMATOGRAPHY APPLIED TO THE STUDY OF PROTEIN ASSOCIATION

Protein association has become an important field of interest because it is related in a living organism to the autoregulation and functional activity of proteins. It is investigated using the transport methods of sedimentation, electrophoresis, and steric-exclusion liquid chromatography.

Steric-exclusion liquid chromatography differs from the other methods in that it can be used at all the other stages of studying protein association and not just for determining the equilibrium constant. In other words, it can be used to ascertain which reacting components are present, how many components are involved, the stoichiometry, and the kinetic constants. In many cases it is possible to do an exhaustive analysis of the association of some proteins simply with steric-exclusion chromatography without enlisting the aid of any other method. The method is based on the relationship between the elution volume and the dimensions of a macromolecule. It is well understood theoretically [332-344], and the equilibrium association constants can be calculated from the way the concentration depends on the elution volume, and the kinetic constants from the variances of the chromatographic fronts.

The need to obtain chromatograms for several concentrations means that significant quantities of protein are consumed. Naturally, microbore columns, which can reduce the protein consumption, have been developed for steric-exclusion chromatography [335-337]. In addition, the study of the relationships between enzymatic activity and protein association requires investigations of how additions of enzyme inhibitors, promotors, or substrates to the eluent affect the association[335, 337]. The low eluent expenditures of microbore columns is a distinct advantage when, as in this case, the eluents are expensive.

If a monomeric protein that is in dynamic equilibrium with oligomeric proteins is analyzed with steric-exclusion chromatography, the components will only be separated if the association reaction is slow. If the reaction is rapid, then there will only be one chromatographic peak, with sharp leading and spread-out trailing boundaries. When the speeds of the separation and the association reaction are comparable, one, two, or three partially separated peaks will be seen, but they will not correspond to different components; and if each of the fractions is reprocessed through the chromatograph, the first chromatogram is reproduced. In order to study association reactions the chromatograph is run in a frontal regime, i.e., a wide zone of the protein in question is fed into the column. Under these conditions a constant composition is established on the plateau, while the elution volume of the centroids for both the leading and trailing boundaries corresponds to the weight-averaged volume of all the reaction components.

We shall now consider the determination of the equilibrium constant and rate constant for the association reaction for the simplest n-mer—monomer system.

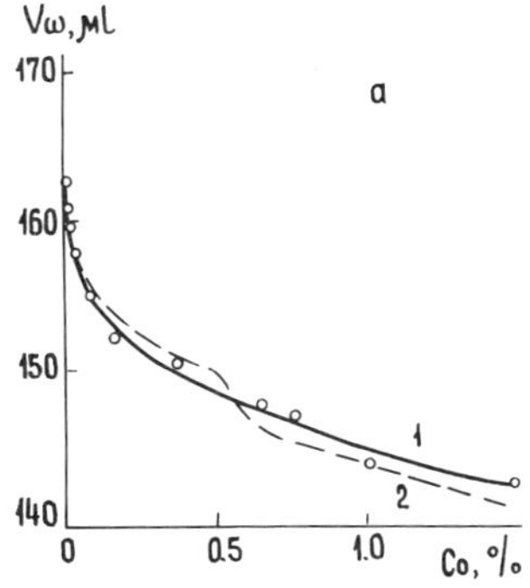

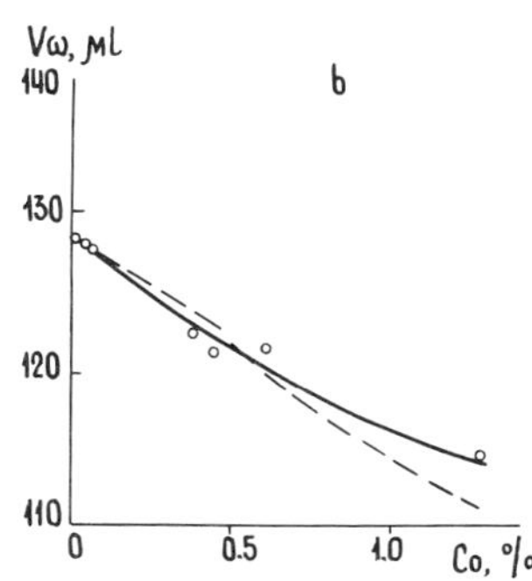

Fig. 6.7. Concentration versus weight-averaged elution volume of lysozyme for a microbore column filled with a) Sephadex G-75 and b) Biogel P-60. 1) Theoretical curve for K = 0.9 dl/g; 2) calculated curve with a doubling of the dimerization constant K. The column was 250 × 1 mm, F = 500 μl/h, and the protein sample was 200 μl [336].

In this case,

$$K_{1,n} = \frac{1 - f_1}{f_1^n C_0^{n-1}}, \tag{6.1}$$

where $K_{1,n}$ is the equilibrium constant, f_1 and f_n are the weight fractions of the monomer and n-mer proteins, respectively, and C_0 is the overall concentration of the protein.

The weight-averaged elution volume $\bar{V}_w = f_1V_1 + (1 - f_1)V_n$ is related to the monomer elution volume V_1, which can be found by extrapolating the $V_w(C_0)$ curve to infinite dilution, and to the elution volume V_n of the n-mer, which can be calculated from V_1 and the calibration relationship $V(M)$ or $V(R_{St})$, where M is the molecular weight of the protein and R_{St} is its Stokes radius.

Thus, the equilibrium constant $K_{1,n}$ and the degree of association n can unambiguously be calculated from the concentrational curve $\bar{V}_w(C_0)$. The calculated and experimental curves $\bar{V}_w(C_0)$ for a microbore steric-exclusion chromatogram of the dimerizing protein lysozyme is shown in Fig. 6.7 [336]; the adsorbents used were Sephadex G-75 and Biogel P-60. The dimerization constant that is obtained from these curves is in good agreement with the results from sedimentation.

The kinetic constants for the association—dissociation reaction can also be ascertained from a microbore steric-exclusion chromatogram. The relationship between the dimensionless variance of

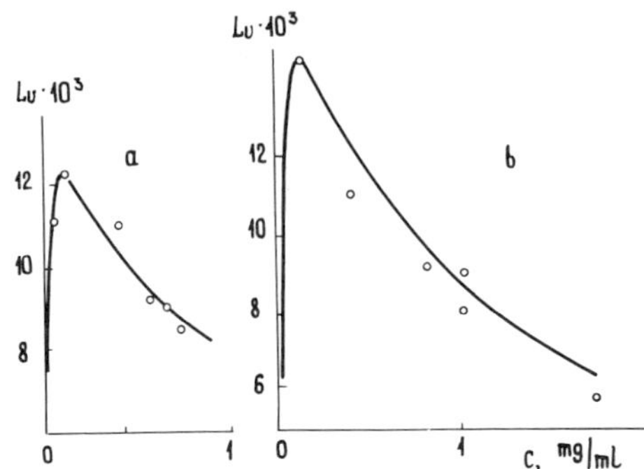

Fig. 6.8. Concentration versus dimensionless dispersion coefficient $L_V(ism) = \sigma_V^2/V^2$ for microbore liquid chromatography of phospholipase A_2 [337] on a microbore column filled with Sephadex G-75. a) No activator; b) with activator (0.02 M Ca^{2+}). The solid line is the calculated curve for $K_2 = 6 \cdot 10^4\ M^{-1}$. In (a) $K_2 = 0.7\ sec^{-1}$, and in (b) $K_2 = 0.2\ sec^{-1}$. Other conditions as in Fig. 6.7.

the leading boundary of the chromatographic zone containing the lipolytic enzyme phospholipase A_2 (L_V) and its concentration is shown in Fig. 6.8 [337]. At low concentrations the protein mainly exists in the monomer form, but at higher concentrations it dimerizes significantly. The spreading of the zone fronts for these limiting cases is solely dependent on the chromatographic process ($L_V = \sigma_V^2/V_R^2 = 1/N$), where N is the number of theoretical plates for the i-th component. However, at intermediate concentrations other factors affect the spreading because, for part of the time, the protein migrates down the column in the monomer form and as the dimer for the rest of the time. This means that the spreading $L_V(C_0)$ passes through a maximum, i.e.,

$$L_V = 1/N + L_{V,k}. \tag{6.2}$$

If the number of theoretical plates the microbore column has for a reference protein is known, then the rate constants for the forward and reverse reactions, k_1 and $k_2 = k_1/K_{1,n}$, can be calculated from the $L_{V,k}(C_0)$ curve [336]. A number of authors [334-338] have demonstrated that the association of proteins can be exhaustively investigated (the equilibrium and kinetic constants determined) using microbore steric-exclusion chromatography with a total protein consumption of less than 1 mg.

6.3. MICROBORE COLUMN CHROMATOGRAPHY OF POLYMERS AND OLIGOMERS

6.3.1. Molecular Weight Distribution of Polymers

The application of microbore steric-exclusion chromatography to the determination of molecular weight distributions is extremely promising, because this method on conventional columns has a number of disadvantages that limit its applicability, such as the inevitability of using considerable quantities of expensive sorbents (e.g., ones with a linear molecular-mass calibration curve) and large volumes of toxic or inflammable solvents for the eluent. The latter limitation excludes solvents such as dimethyl sulfoxide, trifluoroacetic acid, or trifluoroethane, which all dissolve a number of polymers very well. As a result, a number of industrially important polymers, such as polyethylene terephthalate and nylon-6, cannot in practice be analyzed with steric-exclusion chromatography.

By applying microbore columns the quantities of sorbent can be limited to less than 50 mg and the solvent to less than 50-100 μl. Thus, any solvent can be used such as very rare or very pure ones, and total antitoxicity or fire prevention measures can be implemented.

Jinno and Nishihara [339] have applied microbore columns packed with TSK-gel G-3000 ($M = 6 \cdot 10^4$ being excluded) to determine the molecular weight of unsaturated polyethers (the condensation products of maleic anhydride, isophthalic acid, and neopentyl glycol). They were able to separate three components within 50 min on a 300 × 1 mm column using tetrahydrofuran. By calibrating the column with standard molecular weight polystyrene they were able to assign the components molecular weights of $6 \cdot 10^4$, $5.1 \cdot 10^3$, and 280 daltons. Jinno and Nishihara noted that a column 0.25 mm in diameter did not separate the polyethers as well as a column 1 mm in diameter.

Various authors [13, 134, 135, 176, 340] have described high-performance microbore steric-exclusion chromatographs for analyzing molecular weight distributions. They compared the separation of mixtures of polystyrene standards with molecular weights ranging from 2100 to $3.67 \cdot 10^5$ daltons carried out on a conventional system and in a microbore system. The results are shown in Fig. 6.9. To achieve the same separations the conventional system needed two 250 × 3 mm columns and two 500 × 3 mm columns, a total of 1.5 m, whereas the microbore system required just one 330 × 0.6 mm column.

The microbore columns reported in [13, 134, 135, 176, 340] were packed with microspheres of Li Chrospher Si-100 and Si-1000 silica gels ($d_p = 7 \pm 2$ μm) in a ratio of 2:3. The size distribution of the original sorbent had been narrowed by elutriation.

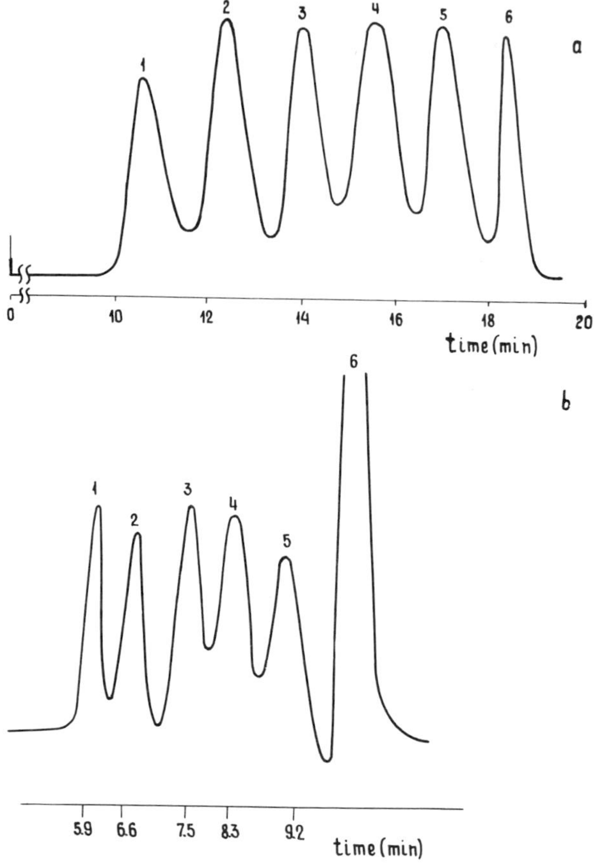

Fig. 6.9. Steric-exclusion liquid chromatography of narrowly fractionated polystyrene standards. a) On a PTFE 330 × 0.6 mm microbore column packed with a mixture of LiChrospher Si-100 and Si-1000. Eluent: dichloromethane at 4.6 μl× min^{-1}. 0.5 μl of sample was injected, containing 2.5 μg of each polystyrene standard. UV detection at 260 nm. b) On three columns: 250 × 3 mm packed with LiChrospher Si-100 (d_p = 10 μm); 250 × 3 mm packed with LiChrospher Si-500 (d_p = 10 μm); and 500 × 3 mm packed with LiChrospher Si-60 (d_p = 5 μm). Eluent: tetrahydrofuran at 1.3 nl/min. UV detection at 254 nm. M_w of polystyrene: 1) 867,000; 2) 200,000; 3) 33,000; 4) 10,000; 5) 2100; 6) benzene. (Reproduced from [134] with the permission of Elsevier Science Publishers.)

The column performance for benzene (k' = 1) came to 7500 theoretical plates with an HETP of 40 μm. Since the retention-volume reproducibility must be better than 0.5% for determining molecular

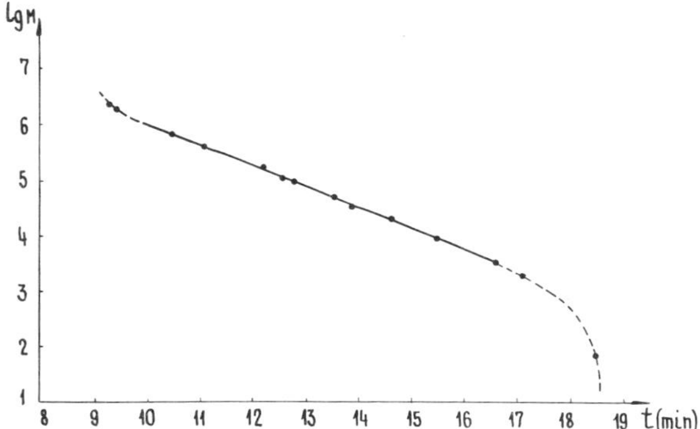

Fig. 6.10. Retention time versus logarithm of molecular weight for the polystyrene standards in Fig. 6.9. 0.3 μg of each standard was introduced in a 0.3 μl sample in dichloromethane [340].

weight distributions, a special syringe pump was prepared. It was made from stainless steel, had a chamber volume of 300 μl and a plunger 4 mm in diameter, and could deliver eluent such that the retention volumes were within 0.1 μl of each other. In order to increase the interpretation reliability of the chromatograms a 330 × 0.6 mm column was used that had an efficiency of 9500 theoretical plates and linear t_R versus log M curves (as calculated in [341]) for molecular weights between $3 \cdot 10^3$ and 10^6. This corresponds, within the limits of experimental error ($\Delta t_R / t_R = 0.5\%$), with the calculated relationship (Fig. 6.10) [341].

In order to obtain the molecular weight parameters for polymer samples, the chromatographic data was interpreted at the fourth level [342], which accounts for spreading by using Pearson distributions. This level of interpretation is necessitated by the considerable asymmetry of the chromatographic peaks, which is apparently caused by asymmetric spreading in the detector cell. Therefore, calibration using higher statistical moments (second, third, and fourth), as shown in Fig. 6.11, other than the calibration shown in Fig. 6.10 were used to interpret the data. The linearity of the t_R–log M relationship is what determined the experimental form of the calibration curves, which have one maximum at the retention time corresponding to the component with $M = 2 \cdot 10^5$ daltons. Table 6.1 gives the M_w and M_n values for a sample of a polydisperse polystyrene standard PS-706 (US National Bureau of Standards) as obtained from the literature and those obtained from microbore steric-exclusion liquid chromatography. The determination of M_w was reproducible to within 5%.

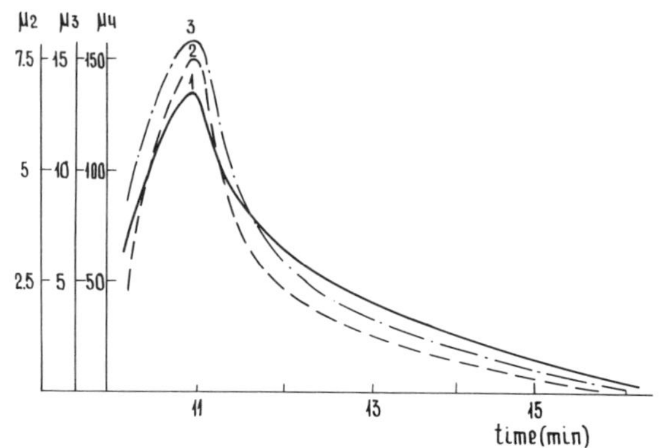

Fig. 6.11. Statistical moment versus retention time for 1) the second, 2) third, and 3) fourth statistical moments of the peaks from a chromatogram of polystyrene standards [340].

TABLE 6.1. Molecular Parameters for the PS-706 Polystyrene Standard

	Steric-exclusion microbore chromatography	Reference [134]
M_n	$(148 \pm 5)\cdot 10^3$	$136.5\cdot 10^3$
M_w	$(273 \pm 5)\cdot 10^3$	$(278 \pm 5)\cdot 10^3$

All these experiments were carried out with a spectrophotometer detector ($V_d < 1$ μl and $\lambda = 260$ nm). However, this sort of detector is of limited use with this technique because only a few polymers absorb ultraviolet light between 250 and 350 nm, which is where the solvents usually used for steric-exclusion chromatography are transparent. Thus, refractive index devices are much more effective for this sort of work.

Aleksandrov et al. [176] have described the use of the LR-1 laser refractometer, which has a cell volume of 0.2 μl (see Section 4.3.3), for separating polymers by microbore steric-exclusion chromatography. A chromatogram of a mixture of polystyrenes obtained from a 165 × 0.5 mm column is shown in Fig. 6.12. The column was packed with a mixture of LiChrospher Si-100 and Si-1000 ($d_p = 7 \pm 2$ μm) and had a linear molecular-weight calibration curve for various elution rates. Notice the asymmetry of the peaks, which increases as the molecular weight goes up or the elution rate increases. However, the asymmetry is less than that obtained when a spectrophotometer is used, given a polystyrene with the same molecular

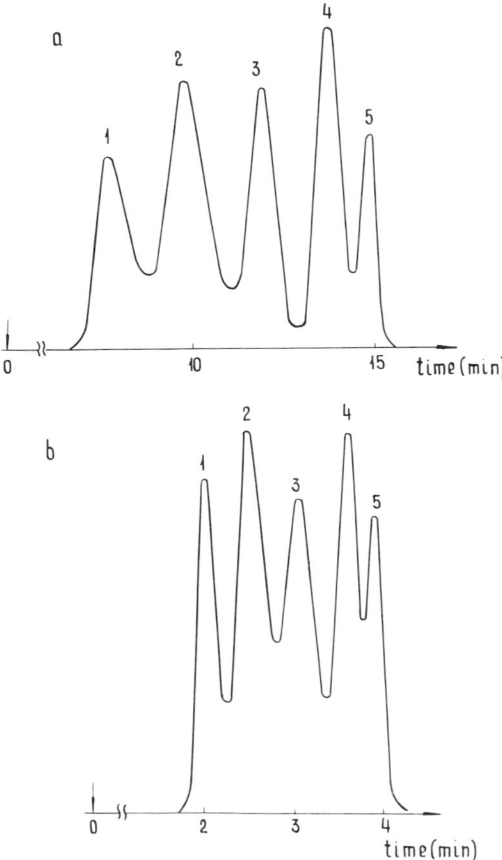

Fig. 6.12. Chromatogram of a polydisperse polystyrene standard taken from a 160 × 0.5 mm PTFE column packed with a mixture of LiChrospher Si-100 and LiChrospher Si-1000. Eluent: dichloromethane at a) 2.3 µl/min and b) 9.2 µl/min. Sample of 0.2 µl contained 0.8 µg of the polystyrene standard in dichloromethane. The detector was an LR-1 laser refractive index device. M_w of polystyrene: 1) 2,300,000; 2) 200,000; 3) 20,000; 4) 2000; 5) benzene. (Reproduced from [176] with the permission of Dr. Alfred Huthig Verlag, GmbH.)

weight distribution and for the same elution rates. By injecting a sample directly into the detector cell it was shown that the peak asymmetry is introduced as the sample flows through the detector and is not due to the column. As mentioned before, this asymmetry means that the chromatographic data has to be corrected [343].

The use of refractive index detectors makes it possible to determine the molecular weight distributions of polyamides and polyethers, for which expensive and toxic solvents are needed.

In 1984 the Science and Technology Organization of the Academy of Sciences of the USSR produced a microbore gel chromatograph, the KhZh-1309. This system contains a microbore column 0.5 mm in diameter for steric-exclusion chromatography, a miniature injector 0.2 μl in volume, and a refractive index detector that has a sensitivity of 10^{-7} RI and a measuring volume of 0.1 μl. The system includes its own computer to calibrate the molecular weights and can calculate the distributions for oligomers and polymers.

6.3.2. High-Performance Microbore Liquid Chromatography of Oligomers

Ishii and Takeuchi [47, 344, 345] have developed ultramicrobore columns for the steric-exclusion and reversed-phase chromatography of oligomers. Their work is a good example of superefficient chromatographic analysis.

They used fused-silica tubes 0.22 and 0.26 mm in diameter and 0.5 m long packed with TSK gel G1000H (5 μm, steric-exclusion limit 10^3) and gel G3000H (5 μl, exclusion limit $6 \cdot 10^4$) for the steric-exclusion chromatography. For the reversed-phase chromatography they used 200 × 0.22 mm and 500 × 0.26 mm columns packed with octadecyl silica gel. The solvents they used for the elution were tetrahydrofuran for the steric-exclusion technique and a mixture of toluene and acetonitrile for the reversed-phase technique. A syringe pump with a chamber of 500 μl provided an elution rate of 0.5-15 μl/min for pressures up to 5.0 MPa; the injector volume was 0.02 μl. The columns were connected together in series to yield an overall column length up to 4 m with an efficiency up to 230,000 theoretical plates ($h \cong 3$) for the steric-exclusion chromatograph, and an overall length of 0.5 m for the reversed-phase chromatograph, which had an efficiency of 30,000-36,000 theoretical plates. The effluent was detected with the UVIDEC-100 UV spectrophotometer (JASCO), which has a measuring volume of 0.04 μl. Chromatograms of the epoxy resin Epicote 1000 obtained by the steric-exclusion technique on columns 2.5 and 4 m long are shown in Fig. 6.13. Ishii and Takeuchi pointed out that although the separations were considerably better than those obtained from conventional columns 1 m long, they took longer. In addition to the main components of the resin, the technique showed up minor peaks, which correspond to isomers of the oligomer homologs [345]. Ishii and Takeuchi also used a superefficient ultramicrobore column with an ODS phase to analyze an aqueous extract of carbon black, which contains very many organic compounds in tiny concentrations, and to separate polycyclic aromatic hydrocarbons and polychlorinated biphenyls.

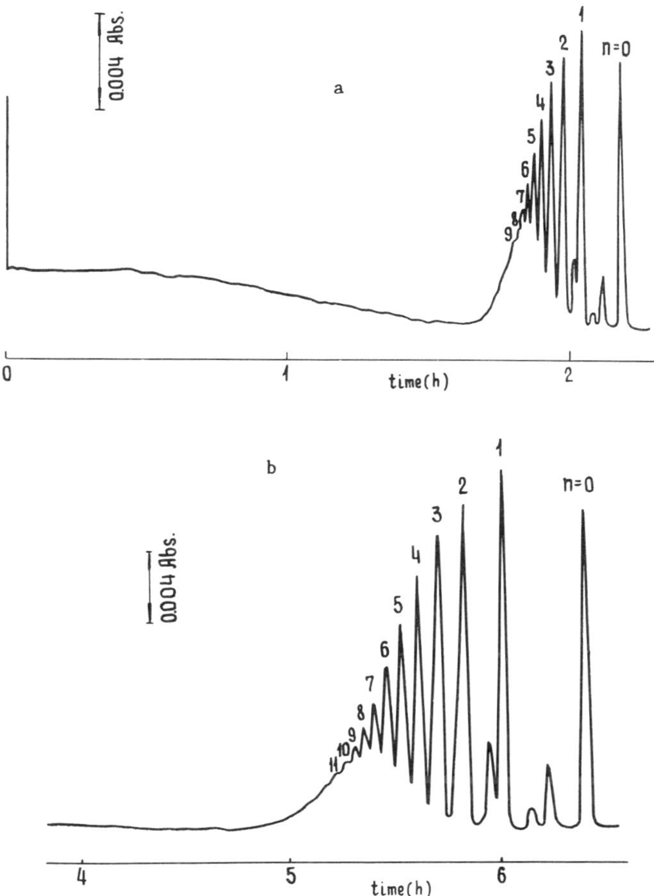

Fig. 6.13. Microbore liquid chromatogram of Epicote 1001 epoxy resin. Taken from a) a 2500 × 0.33 mm and b) a 4000 × 0.33 mm column packed with TSK-Gel G3000H in tetrahydrofuran. The flow rates were a) 1.04 and b) 0.56 µl/min. The sample introduced was 0.02 µl of 3.7% solution. UV detection at 280 nm. (Reproduced from [344] with the permission of Elsevier Science Publishers.)

6.4. ANALYSES OF ANTIBIOTICS AND PHARMACEUTICALS

6.4.1. Antibiotics

Antibiotics have been analyzed with microbore columns packed with silica gels and with reversed-phase microbore columns [346, 347]. Spectrophotometric detection at 110, 254, and 340 nm was used. Neomycin B and C have been separated on a 500 × 1 mm column packed with silica gel spheres (10 μm) using a 590:400:8:2 mixture of chloroform, tetrahydrofuran, water, and glacial acetic acid as the eluent. The analysis took 12 min at an elution rate of 200 μl× min^{-1}. A six-component mixture of novobiocin and its dehydration products was separated using the same column in 12 min. The eluent was a 44:45:5:4:3 mixture of saturated aqueous butyl chloride, butyl chloride, tetrahydrofuran, methanol, and glacial acetic acid, with an elution rate of 300 μl/min. The degradation products of ampicillin have been investigated with reversed-phase chromatography on a 250 × 1 mm column packed with Partisil-10 ODS and eluted with a 115:8:10 mixture of acetonitrile, water, and 2 M ammonium acetate (pH 6). The analysis lasted 52 min with an elution rate of 20 μl× min^{-1}, for which the column performance was 36,800 plates.

Five cephalosporin antibiotics were separated within 30 min on a 250 × 1 mm column packed with μ-Bondapak C_{18} (10 μm) using isocratic elution by a 75:25 mixture of aqueous 0.01 M NaH_2PO_4 and methanol with step changes in the rate from 50 μl/min to 150 μl/min for 30 min. The sample was 5 μl in volume and contained 0.05-0.2 μg of antibiotic. A method has been used to analyze the fermentation broth from an enzyme reactor without preprocessing. The sample was simply diluted with ten times as much water, and after settling it was diluted further in 50 times as much water. The amounts of acetylcephalosporin C and cephalosporin C, which are in quantities of a few micrograms, can then be quantitatively and accurately estimated, the relative standard deviation being 0.75%. The analysis can be accomplished within 6 min if the elution rate is increased to 300 μl/min (pressure ΔP = 20 MPa). Tsuji and Binns [347] pointed out that in order to analyze cephalospirin within 6 min on a conventional column it would require an elution rate of 6 ml/min and a pressure of 35 MPa. Programming step eluent changes with 10% increases in the methanol concentration has turned out to be an effective method for determining the low-level impurities in cephalosporin antibiotics, while keeping the baseline oscillations to a minimum. Tsuji and Binns attributed the smallness of the oscillations to the shortness of their detector (1 mm instead of 10 mm). This reduced the liquid lens effect when measuring the changes in the refractive index due to changes in the eluent composition. Hence, false chromatographic peaks, which would prevent the detection of impurities, were eliminated. The authors pointed out the stability of the method they developed, which enabled them to reproduce the retention times to between 0.3 and 0.7%, the peak areas to between 1.9 and 9.6%, and the peak heights to between 1.4 and 5.2%. Thus, the determination sensitivity of their method is 16-fold greater than that of conventional high-performance liquid chromatography.

6.4.2. Pharmaceuticals

Tsuji and Binns [347] have described this application. The chromatogram in Fig. 6.14 is of seven steroids and was taken with a

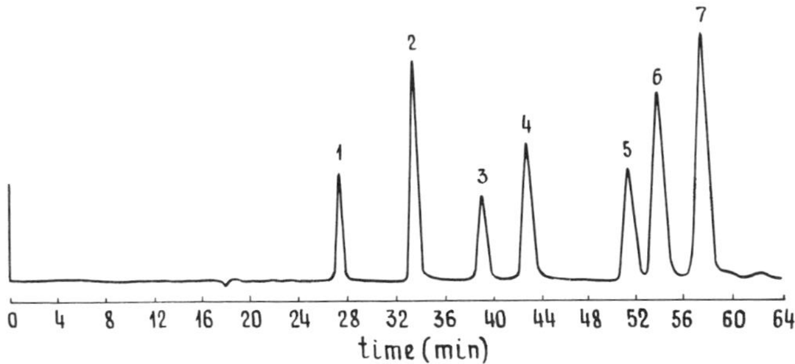

Fig. 6.14. Microbore liquid chromatogram. 1) Fluorometholone; 2) hydrocortisone; 3) hydrocortisone hemisuccinate; 4) methylprednisolone; 5) prednisolone; 6) prednisone. (Reproduced from [347] with the permission of Elsevier Science Publishers.)

500 × 1 mm column packed with silica gel. The eluent was a 450:450:105:58:44 mixture of saturated aqueous butyl chloride, butyl chloride, tetrahydrofuran, methanol, and glacial acetic acid. The maximum column performance was achieved for an elution rate of 5 μl/min and was 60,000 plates. The seven-component mixture was completely separated to the baseline within 60 min. Using a rapid analysis regime and an elution rate of 300 μl/min, a quantitative determination of methylprednisolone, using prednisolone as an internal standard, was accomplished within 3.5 min and for a relative standard deviation of 1.14%. A sulfonylurea preparation (tolbutamide and torazamide) was analyzed within 12 min on the same column using a 475:475:20:15:9 mixture of saturated aqueous hexane, tetrahydrofuran, ethanol, and glacial acetic acid at a rate of 300 $\mu l \times min^{-1}$. A 250 × 1 mm ODS column was used to study ibuprofen with an internal standard of valerophenone using a 65:35 mixture of acetonitrile and water, with 1% chloroacetic acid. The elution rate was 3 μl/min, the pH 3, and the analysis time 8 min.

Fujii et al. [348, 349] used reversed-phase microbore chromatography with spectrophotometric detection to separate and analyze qualitatively the cardiac glycosides contained in the leaves of foxgloves (<u>Digitalis purpurea</u>). They separated the standard five glycosides on a 165 × 0.5 mm column packed with the ODS silica gel SC-01 (5 μm) using a 1:1:1 mixture of acetonitrile, methanol, and water. Given an elution rate of 4 μl/min, the analysis took 35 min. Digitoxin, gitoxin, and their metabolites were separated on a 151 × 0.5 mm column, eluting with aqueous methanol. A linear relationship between the peak area and the mass of the sample was obtained down to a limit of 5-40 ng. It was shown that an accurate quantitative analysis of a 5-ng sample was possible for these compounds.

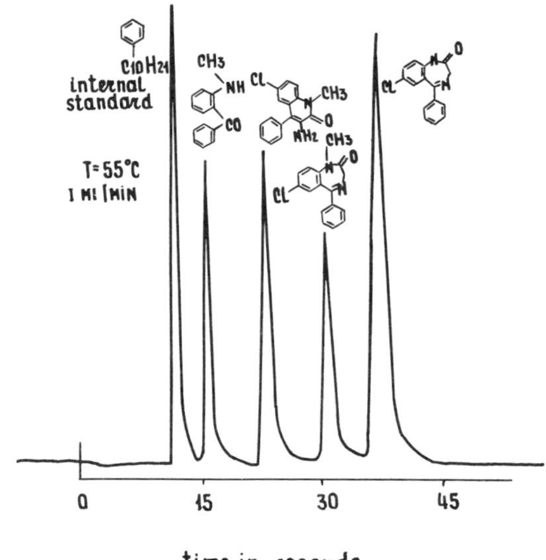

Fig. 6.15. Chromatogram of diazepam and its metabolites taken from a 150 × 0.1 mm column packed with Partisil-10 (d_p = 10 µm). Eluent: methanol, ethyl acetate, heptane (8:10:32) at 1.5 ml× min^{-1}. UV detection at 254 nm. (Reproduced from [54] with the permission of Elsevier Science Publishers.)

Fujii et al. [349] have developed a method for qualitatively estimating the amount of digitoxin and gitoxin in extracts of dried foxglove leaves using the reversed-phase technique in 162 × 0.5 mm columns packed with ODS and with a 15:15:19 mixture of acetonitrile, methanol, and water. The elution flow rate was 4 µl/min and the analysis time 1 h. It was established that 100 mg of dried foxglove leaves contain 0.082 µg of gitoxin and 0.145 µg of digitoxin.

A rapid microbore regime was used by Scott et al. [54] for analyzing diazepam and its metabolites. The diazepam, its three metabolites, and an internal standard were completely separated within 43 sec (Fig. 6.15), the column performance reaching 1600 plates. Kucera and Manius [16] reported the separation of diazepam and its deuterium analog using superefficient microbore chromatography. They used a 450 × 1 mm column packed with ODS silica gel whose performance for benzene (k' = 0.99) was 23,000 plates, the asymmetry factor of the peaks being 1.01. The diazepams were not completely separated because of the formation of peak tails. Interestingly, as it would have done on an unmodified silica gel, the heavier D-diazepam was the first sample to be eluted, even though we would have expected it to have followed the H-diazepam with the reversed-phase technique because the D-molecule should be more hydrophobic. Kucera and Manius believe that the D-diazepam was eluted first because of the relatively high methanol concentration in the eluent and that at the surface of the reverse phase it became like a normal phase.

Kucera and Hartwick [350] described the separation of benzodiazepams on a 500 × 1 mm column packed with silica gel (d_p = 10 μm). The eluent was a 5:10:85 mixture of methanol, ethyl acetate, and heptane, and the column efficiency $2.5 \cdot 10^4$ plates for a compound with k' = 6 and a linear elution velocity of 0.03 cm/sec. They also separated quinine and quinidine on a 250 × 1 mm column packed with Whatman Partisil-10 ODS-3 using a mixture of 15% methanol and 0.01 M KH_2PO_4. Microbore columns were much more sensitive for quinine and quinidine than conventional columns. Using the microbore technique, the sample size was 6-7 pmoles, while a conventional column requires 30 pmoles, the threshold masses being 77 and 305 fmoles, respectively. Either way the analysis requires about 6 min, the detection being done with the fluorescence detector made by Farrand.

Hartwick and Dezaro [58] have developed a rapid microbore liquid chromatograph for the analgesics phenacetin, methylparaben, aspirin, and caffeine. They used a 100 × 1 mm column packed with a C_{18} sorbent (3 μm) and a mixture of 18% acetonitrile and 0.005 M sodium octylsulfate. These compounds, together with two impurities, were separated in 20 sec.

6.4.3. Vitamins

Kucera and Manius [16] have demonstrated that it is possible to apply microbore columns for the analysis of retinol (vitamin A), retinoids, and carotenes. They separated an artificial mixture of seven standards and vitamin derivatives on a 500 × 1 mm column packed with Partisil-10 (10 μm) and using a 2:10:88 mixture of methanol, ethyl acetate, and heptane at an elution rate of 25 μl/min. The analysis was completed down to the baseline within 30 min. A very rapid result was obtained for the separation of tocopherol and trans-retinol on a 150 × 1 mm column [58]. It was packed with Partisil-10 and eluted with chloroform. The vitamins required a total of 4 sec to be separated, given an eluent velocity of 8 cm/sec.

6.5. ANALYSES OF NUCLEIC ACID COMPONENTS

Ion-exchange [351, 352] and reversed-phase [350] microbore liquid chromatographies have been used to analyze the bases, nucleosides, and nucleotides in nucleic acids.

Rokushika et al. [351] separated a mixture of 4 bases, 4 nucleosides, and 12 nucleotides (21.5-26 pmoles of each) using an anion-exchange microbore column. The 450 × 0.19 mm column was made from fused silica and packed with TSK-gel-IC-anion-PW. The mobile phase was 30 mM phosphate buffer, pH 7.15, and the analysis took 1 h. Wulfson and Yakimov [352] have developed an automatic microbore system for splitting oligonucleotides enzymatically and then separating the products on a chromatograph. The enzyme was phosphodiesterase immobilized on cellulose and kept in a 25 × 0.5 mm column. The chromatograms were taken from a series of connected 100 × 0.5 mm columns packed with Aminex A-27 ion-exchange resin. A mixture of nucleosides and mononucleotides was separated using gradient elution in an acetate buffer, pH 4.3, for 2 h. Given an efficiency of 300-500 plates, some 10^{-9} mole of oligonucleotide was detected. Hence, as Wulfson and Yakimov pointed out, large improvements in the analysis sensitivity are possible by increasing the efficiency of the column and the sensitivity of the detector.

Kucera and Hartwick [350] have developed a reversed-phase method on 250 × 1.0 mm columns packed with Partisil-10 ODS. They have used it to separate deoxyribonucleosides and their mononucleotides, and the dimers of ribonucleotide monophosphates. The eluent employed was 0.01 M KH_2PO_4, pH 5.6, with 20% methanol, and the flow rate 50-75 µl/min. A sample of 2 pg was introduced. This technique has been applied to the analysis of modified nucleosides in urine, blood, and tissue cultures, in which they are present in very low concentrations.

6.6. ANALYSES OF MEDICINES AND BIOLOGICALLY ACTIVE COMPOUNDS

When analyzing medicines and biologically active compounds, their concentrations and the detection sensitivities of microbore technology are usually sufficient for a direct analysis. However, for trace analyses, low molecular extinctions, or electrochemical investigations, the sample may have to be concentrated or a correspondingly more sensitive detector applied.

6.6.1. Direct Microbore Analysis of Biologically Active Compounds and Pharmaceuticals in Biological Fluids

6.6.1a. Eicosanoids in Tissue Extracts. Raydrick et al. [353] have reported the use of a reversed-phase 250 × 2 mm column packed with Ultrasphere ODS and UV detection to analyze the eicosanoids in extracts from the medulla tissue of kidneys. Figure 6.16 shows the separation of 11 standard eicosanoids in a 69:31 mixture of 0.0025 M orthophosphoric acid and acetonitrile; the separation took 35 min.

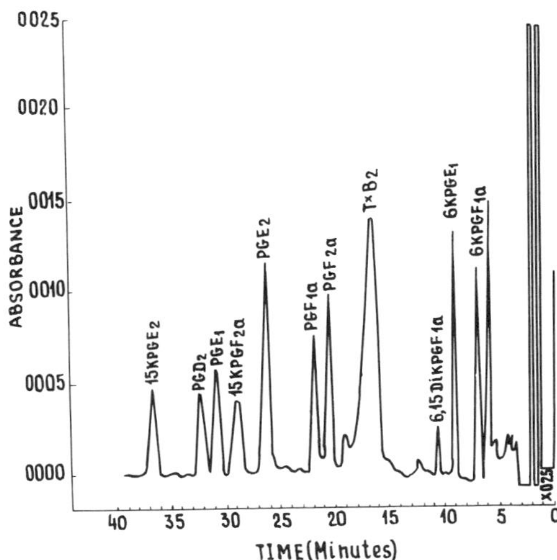

Fig. 6.16. Microbore liquid chromatogram of a mixture of eicosanoids. Eluent flow rate of 0.3 ml/min, and 2 ng of each standard were introduced. UV detection at 190 nm. (Reproduced from [355] with the permission of Elsevier Science Publishers.)

The threshold sample mass for this analysis was 250 pg for the standard solution. The eicosanoids were not identified by their structure (e.g., using a mass spectrometer), but indirectly a) by comparing the retention times of the components with those of standard solutions, b) by comparing the height of an absorption peak and its concentration, c) by measuring the radioactivity of the peak when 1-[^{14}C]arachidonic acid is used as the substrate, or d) by the inhibition with meclofenamic acid.

6.6.1b. Bethanidine in Blood Plasma. Shipe et al. [354] used a 250 × 2.1 mm column to determine the bethanidine content of blood plasma. They used the reversed-phase technique with C_{18} sorbent (5 μm) with an internal standard of N-methylphenylethylamine. It took 15 min to obtain a quantitative analysis using an elution rate of 0.25 ml/min with isocratic elution of a 9:1 mixture of a buffer acetate and acetonitrile, with some added tetramethylammonium chloride. The ratio of the height of the bethanidine peak to that of the internal standard was found to be linearly dependent on the quantity of bethanidine in the plasma for bethanidine contents of between 0.7 and 5 mg/liter, with a correlation coefficient of 0.996. The method enabled Shipe et al. to study how the concentration of bethanidine in the blood plasma of a patient was a function of the elapsed time after the injection of 20 mg/kg of the medicine. They noted that by using a 2.1-mm column instead of a 4.6-mm column they obtained, for the same experimental conditions, fivefold larger peak areas and significantly less noise. The method could thus be used for a routine analysis of the drug level in the plasma. They also reported that the microbore column could be used for analyzing 500 plasma samples.

6.6.1c. Rapid Analyses of Theophylline in Blood Serum. In order to separate theophylline from theobromine and similar components of blood plasma, Kucera and Hartwick [350] employed a high-speed microbore analysis (analysis times of 5-10 sec). They used 1-mm columns 360-600 mm long packed with a reversed-phase sorbent and a linear elution rate of 1-4 cm/sec. This optimization with respect to particle dimension, elution rate, and column length, etc., yielded analysis times from 40 to 100 times faster than those possible on standard columns, i.e., columns 100-250 mm long by 46 mm wide.

6.6.1d. Tyrosine and Tryptophan Metabolites in Urine. Hirata et al. [189] have described a technique for finding the neutral and acidic metabolites of tyrosine and tryptophan in urine using packed capillary columns and an electrochemical detector ($V_d \sim 0.15$ µl). The columns used for the separation were packed with Merck's Li Chrosorb Si-1000 silica gel ($d_p = 30$ µm); the injector volume was 0.1-1 µl. A standard mixture of methoxyhydroxyphenyl glycol, p-hydroxyphenylacetic acid, 5-hydroxyindoleacetic acid, homovanillinic acid, and vanillinic acid each in a concentration of $1.6 \cdot 10^{-4}$ M was completely separated in 8 h. When 1 µl of acidic human urine was injected into the column, all but one of the above four compounds were detected. The exception was the methoxyhydroxyphenyl glycol, whose detection was hindered by the fall in column efficiency due to the increase of the sample volume from 0.3 µl to 1 µl. The threshold mass of the sample was 0.3-1 ng. Hirata et al. emphasized that the high performance of the column and the sensitivity and selectivity of the electrochemical detector (it was much more sensitive than either UV or fluorometer detectors for these compounds) yields a technique that is particularly useful for analyzing electroactive species in complicated biological liquids. Moreover, the preparations do not have to be preprocessed.

6.6.2. Ultrasensitive Microbore Analyses of Concentrated Samples

The determination of trace amounts in biological liquids, such as urine or blood, raises the problems of concentration and preprocessing to eliminate more concentrated solutions that are not of interest.

As we showed in Section 3.3.1, either the sample volume must be a few percent of the retention volume in order not to reduce the column performance, or if the sample volume is large, then the sample must be delivered in a solvent for which the compound we are interested in will be sorbed in a narrow zone at the top of the column. Since the latter approach is limited, we must first concentrate the sample and rid it of the more concentrated components.

Several systems of connecting the precolumns and analytical microbore chromatograph have been developed. The simplest is an off-line system, in which the sample is concentrated, cleaned of the uninteresting impurities, and dried in the precolumn before being introduced into the analytical system. An improvement is the on-line method, in which stationary precolumns are switched into and out of the system either manually or automatically using a system of taps.

6.6.2a. Analyses of Corticosteroids in Blood Serum Using Off-Line Precolumns. Ishii et al. [41] described this technique for microbore columns. The sample was concentrated in off-line precolumns packed with a nonpolar porous polymer sorbent, the columns being regenerated by washing them with methanol and water. Note that this method of sample feed does not reduce the analytical column's efficiency or resolution, nor does it alter the retention times. They eventually chose Hitachi gel-3010 as the sorbent in the precolumn, as it ensured the quantitative sorption and desorption of the solute.

The sample was taken from 2 ml of blood serum that had been diluted tenfold. The subsidiary column was then washed with 0.2 ml of water (increasing the wash volume to 8 ml did not change the peak heights, the retention times, or the column performance or resolution), drained, and the working column switched in. The eluent was taken from the precolumn.

A flow-type spectrophotometer with measuring cells from 0.05 to 0.4 μl in size served as the detector, and the quantity of corticosteroid in the sample was determined by comparison with a calibration curve obtained using a test mixture. The threshold concentrations of the steroids in a sample containing 0.2 ml of rabbit blood serum and 1.8 ml of water were 1.2, 1.1, and $1.8 \cdot 10^{-4}$% for corticosterone, cortisone, and hydrocortisone, respectively.

Using this method on 0.2-ml samples, Ishii et al. were able to measure 10 ng of corticosterone in rabbit blood serum, and 13 ng of cortisone and 4 ng of hydrocortisone in horse blood serum. The determination accuracy was around 15% and the yield 90-100%.

6.6.2b. Analyses of Catecholamines in Biological Fluids Using Automatic On-Line Concentration. The system included a 20 × 0.5 mm microbore precolumn packed with alumina (d_p = 5 μm) to concentrate the urine components before the main column [196] (see Section 4.2.4). A test mixture of catecholamines demonstrated that up to 40 ng of

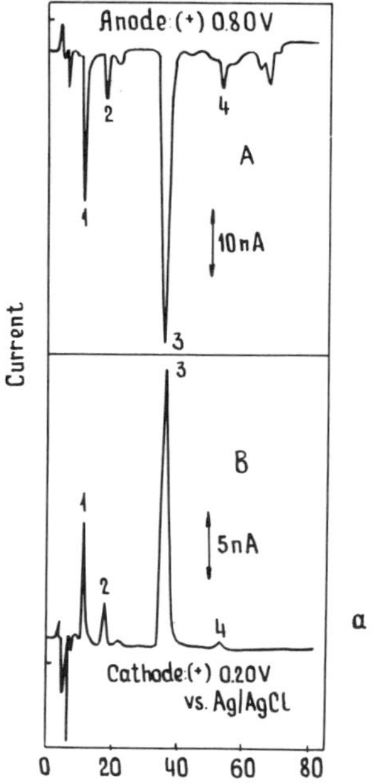

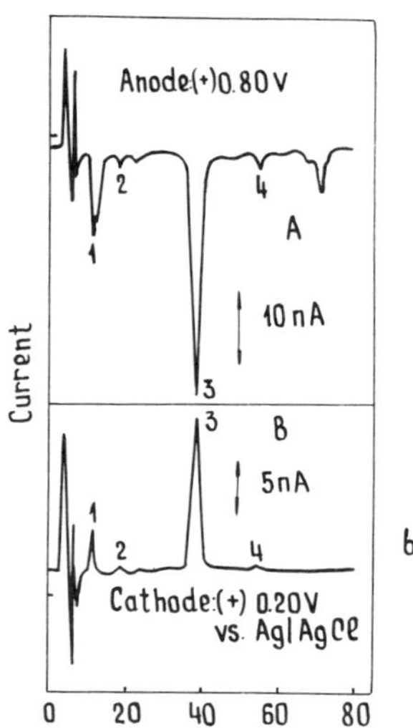

Fig. 6.17. Trace analysis for catecholamines in 100 μl of human urine. Chromatogram taken on a 150 × 0.5 mm column packed with ODS sorbent (d_p = 5 μm). Initial concentration of sample on a 20 × 0.5 mm column packed with alumina. Eluent: Britton—Robinson buffer, pH 1.8, with 0.5 mM of the sodium salt of heptanesulfonic acid. Flow rate: 8.3 μl/min. Dual electrochemical detector (V—Ag/AgCl): anode voltage +0.80 V; cathode voltage +0.20 V. 1) Noradrenaline; 2) adrenaline; 3) dopamine; 4) ℓ-dopa. A) Anode chromatogram; B) cathode chromatogram. (Reproduced from [196] with the permission of Elsevier Science Publishers.)

them could be sorbed quantitatively in the precolumn. By using a dual electrochemical detector the reversible electrochemical reaction of the catecholamines could be selectively detected against the irreversible background reactions of the other components of urine. The catecholamine cathode and anode signals were linearly dependent on the size of the input sample. The anode and cathode signals for a chromatogram of 0.1 ml of human urine are given in Fig. 6.17. Four catecholamines were completely separated on a 150 × 0.5 mm column packed with ODS silica gel (d_p = 5 μm) within 1 h. Note the shoulder on the adrenaline peak and the small extra peak after the exit of ℓ-dopa in the anode chromatogram. The cathode chromatogram does not have these components because they are irreversibly oxidized at the anode. It is possible that the shoulder on the adrenaline peak is due to one of the catecholamine me-

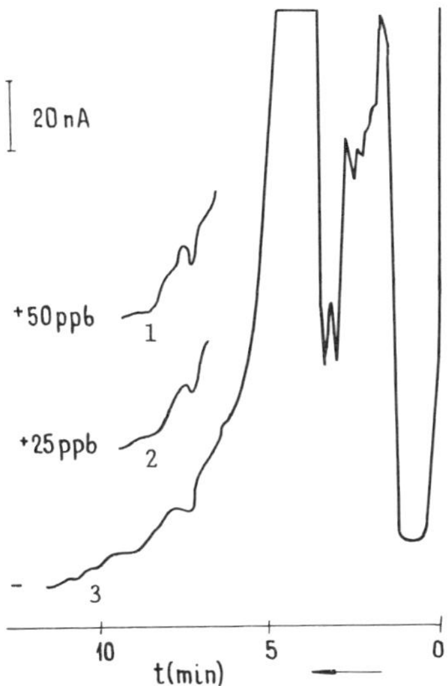

Fig. 6.18. Trace analysis of 4-nitrophenol in urine. Chromatogram taken on a 250 × 1.1 mm column packed with LiChrosorb RP-18 (d_p = 7.7 μm). Eluent: 1:1 mixture of methanol and water plus 10 mM potassium nitrate and 1 mM nitric acid, flow rate 100 μl/min. Sample concentrated beforehand on a 35 × 1.1 mm precolumn packed with LiChrosorb RP-18 (d_p = 7 μm). Samples of 100 μl introduced containing 1) $50 \cdot 10^{-7}$%, 2) $25 \cdot 10^{-7}$%, and 3) no 4-nitrophenol. Electrochemical detection with E = −900 mV. (Reproduced from [165] with the permission of Elsevier Science Publishers.)

tabolites. The method will analyze 30-60 ng of catecholamines with standard deviations of 0.6% for noradrenaline, 0.9% for adrenaline, 1.8% for dopamine, and 5.1% for ℓ-dopa.

6.6.2c. Determination of Nitrophenol in Urine. Kok et al. [165] have suggested a method for concentrating samples so as to detect and measure trace solutes contained in biological fluids and without the need for any special equipment. It uses the on-line technique described in Section 4.4, and involves a 35 × 1.1 mm precolumn packed with LiChrosorb RP-18 (d_p = 7 μm). In principle, it is possible to use a precolumn only 10 mm long, which would give a 90% yield of a compound with a k' ≧ 2 in a 50:50 water–methanol eluent; moreover, the concentration method should not affect the performance of the analytical column. However, after processing several samples using this setup the performance of the analytical column falls from 5000 to 3500 plates. The column performance does not then change even if operated for several weeks.

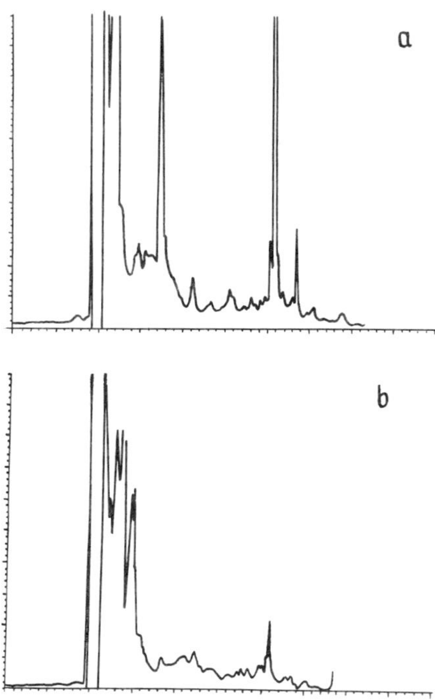

Fig. 6.19. Trace analysis of blood from a) a healthy person and b) a sick patient. Chromatograms taken on a 1000 × 1 mm column packed with RP-18 (d_p = 10 μm) using an exponential gradient over 45 min from a 75:25 methanol–water system to pure methanol at a flow rate of 50 μl/min. Concentration in a dual loop injector. UV detection at 254 nm. Sample of 400 μl of blood. (Reproduced from [53] with the permission of Elsevier Science Publishers.)

Figure 6.18 contains a chromatogram for the 4-nitrophenol contained in 0.1 ml of urine. The limiting detectable concentration with an electrochemical detector is $2.5 \cdot 10^{-6}$%, but the figure shows that the concentration of the 4-nitrophenol is below this value. The determination accuracy with respect to peak height for concentrations between 5 and $20 \cdot 10^{-6}$% was ±10%.

6.6.2d. Blood Serum Analyses Using Sample Concentration and Serial Loop Injectors. Scott and Kucera [53] have suggested a method of concentrating blood serum by using two serially connected loop injectors (Fig. 4.33). The first injector takes the sample, while the second concentrates it. To accomplish this a column packed with a silica gel or hydrophobic sorbent with grain dimensions of 100-120 μm is established in place of an injection loop. The sample is moved from the loop to the column by the pump which delivers the solvent for replacing the sample and washing the concentration column. It is important for the eluent carrying the sample from the concentrating column to the working one to have a lower eluting strength in the working column than in the concentrating one. Thus, the concentrating column is packed with a sorbent whose capacity is smaller than that in the working column. For example, the concentrating column can be packed with a surface-porous sorbent and the working column with a volume-porous sorbent, the active group modification for each being the same. Alternatively, the alkyl groups in the concentrating column's alkylsilica gel can be made shorter than those in the working column (C_2 versus C_8, or C_8 versus C_{18}).

The construction of such a system is considered in more detail in Section 4.4. The approach has been applied to a system for testing 0.4 ml of human blood serum, and the chromatograms for sick and healthy people (Fig. 6.19) can be seen to differ markedly.

6.6.3. Analyses Using Precolumn Derivatization for Medicines and Their Metabolites

Derivatization or the creation of compounds that absorb strongly in the ultraviolet or visible spectrum from the species we are interested in by reacting them with special reagents can raise the sensitivity and selectivity of detection. Even greater sensitivity gains can be achieved by creating fluorescent derivatives. The derivation reaction can be carried out either before or after the column (i.e., precolumn or postcolumn derivatization). Since postcolumn derivatization raises the extracolumn spreading, precolumn treatment is preferable. However, it is only possible when the derivative of the component we are interested in is unique to it and is easily separated in a chromatograph.

6.6.3a. Glycosides and Their Metabolites in Blood and Urine. Fujii et al. [355] used precolumn derivatization with 3,5-dinitrobenzoyl chloride for determining the cardiac glycosides and their metabolites in blood and urine. The formation of these derivatives increases the detectability of the solute in question. Five 3,5-dinitrobenzoyl derivatives of digitoxin and six derivatives of β-methyldigoxin and their metabolites were separated on a 150 × 0.5 mm column packed with C_{18} silica gel (5 μm) using a 3:1:1 mixture of acetonitrile, methanol, and water at an elution rate of 8 μl/min. A spectrophotometer tuned to 230 nm provided a linear relationship between peak area and concentration for quantities of 1.3-35 pmoles. The accurate detection limits were 2, 1, 2, and 0.4 ng for the digitoxin, digitoxigenin, digoxin, and digoxigenin derivatives, respectively. An internal standard of gitoxin was needed for determining these compounds in urine and blood.

6.6.3b. Other Investigations. Novotny et al. [356] used reversed-phase packed capillary columns to determine the steroid me-

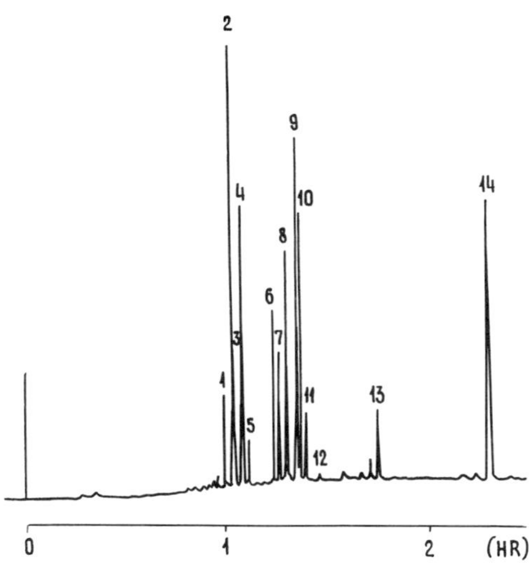

Fig. 6.20. Reversed-phase microbore liquid chromatogram of a standard mixture of the benzoyl derivatives of steroids. Step elution of acetonitrile in water: 80% for 15 min, 85% for 14 min, 90% for 15 min, 95% for 18 min, and 100% to end; flow rate 1 μl/min; UV detection at 230 nm. Steroid derivatives: 1) 11-hydroxyandrosterone; 2) 11-hydroxyetiocholanolone; 3) allo-tetrahydrocortisol; 4) tetrahydrocortisol; 5) tetrahydrocortisone; 6) β-cortolone; 7) β-cortol; 8) α-cortolone; 9) α-cortol; 10) etiocholanolone; 11) androsterone; 12) fehydroepiandrosterone; 13) pregnanetriol; 14) androstanediol. (Reproduced from [359] with the permission of the American Chemical Society.)

tabolite contents in the urine of diabetic and normal humans. In order to raise the sensitivity of detection they analyzed the compounds' benzoyl derivatives. A chromatogram obtained from a 1000 × 0.24 mm capillary column packed with ODS silica gel spheres (3 μm) is given in Fig. 6.20. It shows the separation of 14 steroid derivatives and took 2.5 h using a stepped gradient regime with an acetonitrile—water eluent. Several of the steroids are so close in structure that they have not been separated by gas chromatography. The method was used to obtain the chromatographic steroid metabolite profiles for diabetic and normal individuals. The threshold sample mass of androsterone in these analyses is 500 pg, compared to 4 ng for a conventional high-performance liquid chromatograph with a column diameter of 4.6 mm.

Precolumn derivatization and packed capillary columns were applied by Novotny et al. [171] to raise the detection sensitivity for steroids, prostaglandins, and bile acids. The steroid conjugates (glucuronide sulfates) in urine were treated with DNS-hydrazine to form a fluorescing derivative. The mixture was separated on a 1000 × 0.25 mm column packed with Spherisorb C_{18} (5 μm) using an isocratic regime with a 65:35:1 mixture of 0.01 M sodium acetate in methanol, water, and glacial acetic acid as the eluent. The separation took 1 h.

They also used a new fluorescing reagent, viz., 7-(chlorocarbonylmethoxy)-4-methylcoumarin, to raise the sensitivity of their system for hydroxy compounds, in particular the "solvolyzable" plasma steroids and prostaglandins. Eleven components were separated on a 1500 × 0.24 mm column packed with 3-μm ODS silica gel operated with an acetonitrile gradient of 75% to 100% in water. The analysis took 3 h. Two prostaglandin derivatives were separated on a 1000 × 0.24 mm column packed with ODS silica gel (5 μm) operated with an acetonitrile gradient of 65% to 100% in water, with 1% acetic acid.

To raise the sensitivity for bile acids in biological liquids, Novotny et al. treated them with bromomethylcoumarin [171]. The separation of nine bile acid derivatives was accomplished on a 1000 × 0.24 mm column packed with 5-μm Spherisorb ODS silica gel operated with a stepped gradient regime of acetonitrile in water.

6.6.4. Postcolumn Derivatization for Bile Acid Assays

A number of authors [170, 224, 225] have described a fused-silica packed capillary system with precolumn concentration and postcolumn derivatization for assaying bile acids. The derivatization involves the enzymatic oxidation of the bile acids' 3α-OH groups to ketones, with the reduction of a cofactor, β-nicotinamide adenine dinucleotide (NAD), to NADH, which is detected by its fluorescence. The enzyme, 3α-hydroxysteroid dehydrogenase, is immobilized by covalent bonds on aminopropyl macroporous glass in a special column reactor placed after the separating column. During the reaction the NAD, which is included in the eluent, is reduced to NADH, which is then detected.

The relative merits of precolumn and postcolumn derivatization, as described in Section 4.2.10, were investigated in [225].

The bile acids were separated on 0.26-mm columns 100-250 mm long and packed with Bilepak or ODS SC-01 silica gel (both 5 μm and made by JASCO). A 10 × 0.2 mm guard column made of PTFE and packed with the same sorbent was placed in front of the separating column. A separation of 15 bile acids, which took 2 h, is shown in Fig. 6.21. It was obtained by reversed-phase chromatography with gradient elution. The eluent systems consisted of acetonitrile, 30 mM

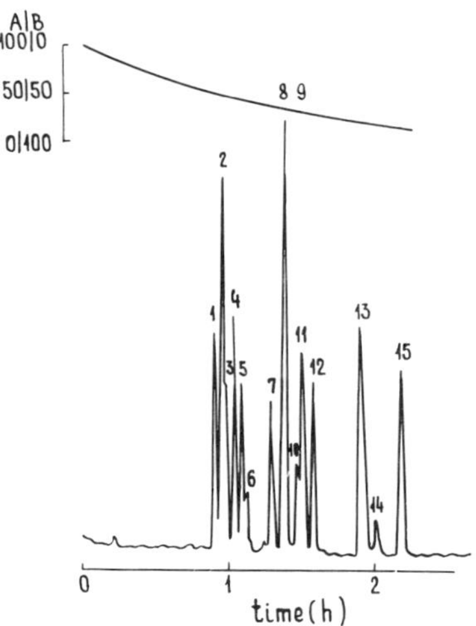

Fig. 6.21. Reversed-phase microbore liquid chromatogram of bile acids with NAD mixed with the eluent in a precolumn, the chromatogram being taken on a 250 × 0.26 mm column packed with Bilepak. The gradient profile is given in the figure, and the flow rate was 1.4 µl/min. Fluorescence detection, with excitation at 365 nm and measured at 470 nm. Sample size: 40-50 ng. (Reproduced from [225] with the permission of Elsevier Science Publishers.)

KH_2PO_4 (pH 7.0), 0.1% ammonium carbonate, and either an 18:52:30 mixture of 6 mM NAD, 0.05% 2-mercaptoethanol, and 1 mM EDTA for eluent A or 35:35:30 mixture of the latter three components for eluent B. The threshold sample mass yielded by the method was 0.09-0.7 ng. The same technique, but using a 200 × 0.26 mm column packed with ODS SC-01 silica gel, was applied to an assay of calf blood serum. The sample was diluted tenfold in the phosphate buffer (pH 7.8) and then concentrated in a precolumn. Only 0.1 ml of serum was needed for the analysis, which took 100 min. Ishii and others emphasized the reproducibility of their results with respect to retention times (relative standard deviations of 1-1.5% when a sample from the precolumn is injected) and peak heights (relative standard deviations of 1.5-2% when there is precolumn concentration). For concentrations between 10 and 40 ng/ml the peak heights are proportional to concentration.

Ishii et al. [170] found that the pHs 8.8-8.9 were optimum and that they simplified the eluent compositions needed for the selective separation of the bile acids.

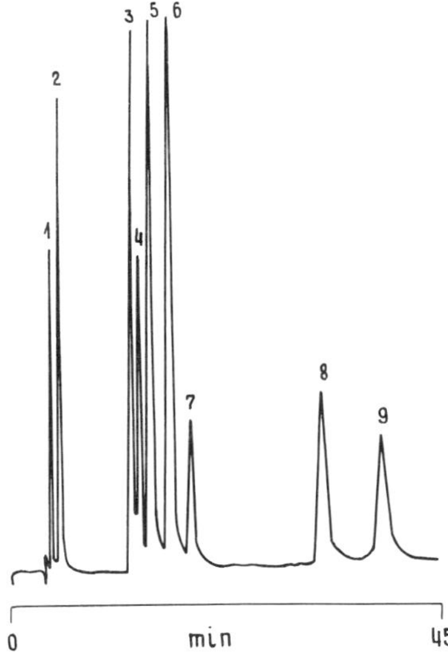

Fig. 6.22. Liquid chromatogram of a standard mixture of nitropolycyclic aromatic hydrocarbons, taken on a 250 × 1 mm column with a pyrene-bound silica gel in hexane. The flow rate was 50 μl/min, and UV detection was used at 254 nm. 1) Naphthalene; 2) pyrene; 3) nitronaphthalene; 4) 4-nitrobiphenyl; 5) 9-nitroanthracene; 6) 3-nitrobiphenyl; 7) 2-nitronaphthalene; 8) 4-nitroterphenyl; 9) 1-nitropyrene. (Reproduced from [360] with the permission of Preston Publications Ltd.)

6.7. MICROBORE LIQUID CHROMATOGRAPHY IN ENVIRONMENTAL MONITORING

In order to be able to monitor environmental pollution, rapid, sensitive, and specific methods for separating and identifying trace quantities of toxic and carcinogenic compounds, such as polycyclic and nitropolycyclic aromatic hydrocarbons and chlorinated phenols, are essential. These tests can all be carried out more efficiently on microbore columns with concentrational detectors than on conventional high-performance liquid chromatographs because a smaller mass of sample has to be preprocessed to concentrate it or to increase the analysis sensitivity.

6.7.1. Analyzing the Content of Polycyclic and Nitropolycyclic Hydrocarbons in Diesel Fuel

Jin and Rappaport [193] have described a method for separating and identifying 12 nitropolycyclic aromatic hydrocarbons using two columns connected in series. Each column was 500 × 1 mm and packed with a reversed-phase sorbent; the elution rate was 40 μl/min. The solute was detected using an electrochemical detector working with an electrode potential of 0.6 V. The method showed the presence in diesel fuel of 1-nitropyrene, 0.45-14.7 ng per milligram fuel.

Hirose et al. [360] used microbore chromatography for analyzing polycyclic and nitropolycyclic aromatic hydrocarbons. They employed silanized silica gel with a stationary phase pyrene-bonded to the gel via siloxane bridges in a 250 × 1 mm column. The pyrene-bonded phase was significantly more selective for nitropolycyclic aromatic hydrocarbons than for polycyclic hydrocarbons without nitro groups. The chromatogram in Fig. 6.22 is for seven nitropolycyclic hydrocarbons, naphthalene, and pyrene in n-hexane; it was obtained within 45 min using an elution rate of 50 μl/min. The column performance for 2-nitronaphthalene (k' = 4.48) was 39,000 plates/m. The threshold sample mass of pyrene was 70 pg (for UV detection at 306 nm). Hirose et al. noted that when the phase was phenyl-bonded, the results were worse because they were less selective for the nitropolycyclic hydrocarbons. A mass spectrometer and volatile eluents were employed for the detection and identification.

6.7.2. Polycyclic Hydrocarbons and Chlorinated Phenols in Water

In order to measure the contents of these compounds in water Slais et al. [264] applied microbore reversed-phase columns and electrochemical detection to preconcentrated samples (1 ml) in a noneluting solvent. The eluent was a 6:4 mixture of acetonitrile and water, together with 0.1 M $NaClO_4$ and 0.001 M $HClO_4$, and the analysis took 50 min (u = 1.4 mm/sec). The threshold concentration of chlorophenol in water was 20-280 ng/liter, i.e., 20-230 ppb.

6.7.3. Polycyclic Aromatic Hydrocarbons in Exhaust Fumes

The fumes from a diesel engine were investigated by Ishii et al. [43], who prepared an aqueous extract of the particles collected from the exhaust pipe. The extract was analyzed on a 30 × 0.5 mm column packed with Chemosorb ODS (3 μm) within 5 min. Since the reference samples for identifying the peaks are themselves toxic and using them would add to atmospheric pollution, Ishii et al. proposed an indirect method that does not require standards. They demonstrated that the retainment (k') of polycyclic aromatic hydrocarbons in reversed-phase liquid chromatography is correlated with the number of double bonds and the number of primary and secondary carbon atoms. If we find out how log k' depends on these values for nontoxic or semitoxic hydrocarbons, then we can predict the k' for their toxic homologs.

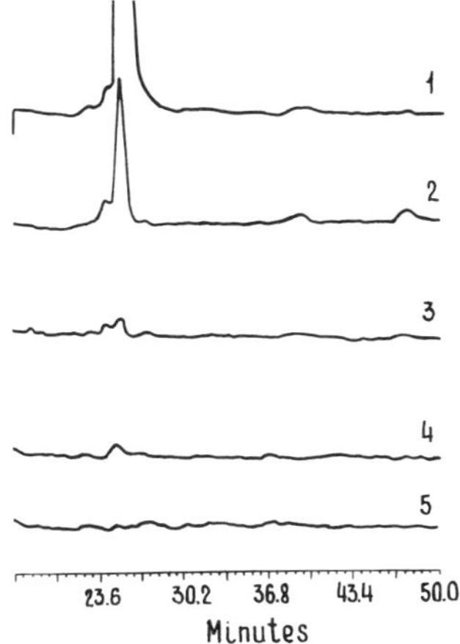

Fig. 6.23. Trace analysis of acetophenone in water using a 500 × 1 mm column packed with ODS-2 (d_p = 10 μm). Eluent: 75:25 mixture of methanol and water, at a flow rate of 40 μl × min^{-1}. Sample concentrated on a precolumn, after being introduced using a twin loop injector. The following acetophenone samples were introduced in 1 μl of water: 1) 10^{-4}%; 2) 10^{-5}%; 3) 10^{-6}%; and in 10 ml of water: 4) 10^{-7}%; and 5) pure water. UV detection at 254 nm. (Reproduced from [53] with the permission of Elsevier Science Publishers.)

Ishii et al. used the approach to calculate the k' for 16 polycyclic aromatic hydrocarbons from the data for naphthalene, biphenyl, anthracene, pyrene, chrysene, and perylene. In a chromatogram of diesel fumes 14 of these compounds can be identified.

6.7.4. Organic Impurities in Water, Given Sample Concentration in Serial Loop Injectors

6.7.4a. General Assay for Organic Compounds. We described the serial loop concentrator in Section 4.4 for the analysis of blood serum. The system can ensure the sorption of aqueous solutions of only slightly polar organic compounds in the form of narrow zones at the start of short precolumns. The analysis sensitivity of the system was tested with aqueous solutions of acetophenone [53]. It turned out that even distilled and deionized water was not pure enough with respect to hydrophobic impurities. To purify the water, it had to be passed through a 250 × 4.6 mm column packed with Partisil ODS-2. The water was then used to dilute acetophenone to concentrations of 10^{-4}, 10^{-5}, 10^{-6}, and 10^{-7}%. Their chromato-

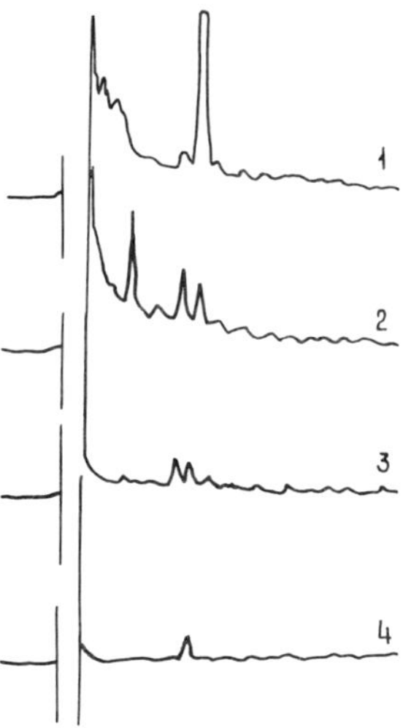

Fig. 6.24. Trace analyses for water from various sources, taken on a 500 × 1 mm column packed with ODS-2 (d_p = 10 μm) using a linear gradient over 1 h from 60:40 methanol–water to pure methanol. 1) River water; 2) mains water; 3) potable water; 4) purified water. Sample concentrated in a twin loop injector system, the remaining conditions as in Fig. 6.23. (Reproduced from [53] with the permission of Elsevier Science Publishers.)

grams are given in Fig. 6.23; the first three concentrations were injected with sample volumes of 1 ml, and the fourth sample was 10 ml. The system's detection limit (UV detector) was 10 ng.

The chromatograms in Fig. 6.24 are for river and mains water (sample size 1 ml). The first two samples clearly have significant quantities of organic compounds, while the third sample has virtually none. We can conclude, on the basis of the threshold acetophenone concentrations for this method, that the potable water contains only 10^{-7}% (1 ppb) organic compounds. The identification of the compounds is not of much interest because there are several kinds of organic pollution of water; the important question is whether the water is polluted.

6.7.4b. Nitrobenzene in River Water. Kok et al. [165] used the on-line technique that we described in Chapters 4 and 6 for determining the nitrophenol content of urine to analyze the nitrobenzene content of river water. A 0.1-ml sample of river water was first passed through a Millipore filter, and KNO_3 and HNO_3 were added to concentrations of 10^{-2} M and 10^{-3} M, respectively. The

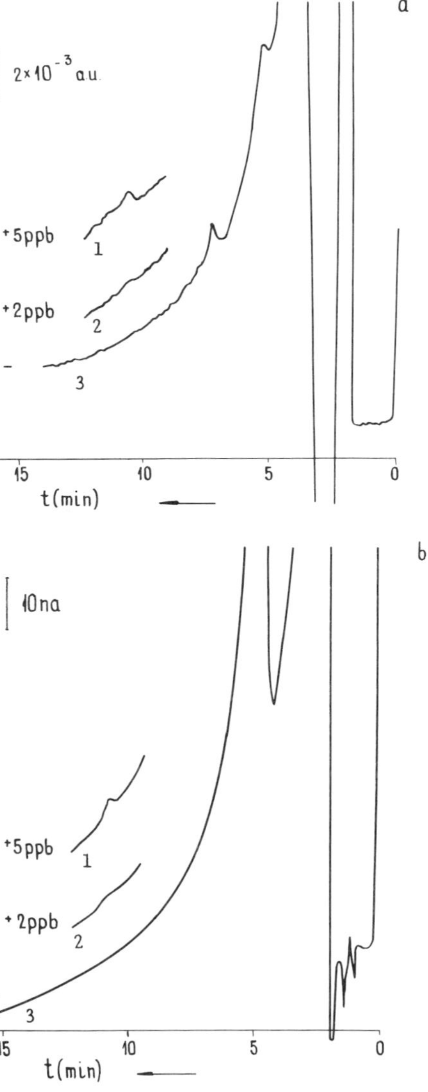

Fig. 6.25. Trace analysis of nitrobenzene in river water. Taken on a 250 × 1.1 mm column packed with LiChrosorb RP-18 (d_p = 7.7 μm). The eluent was a 1:1 methanol—water system that included 10 mM potassium nitrate and 1 mM nitric acid; it was used at 80 μl/min. The sample was concentrated beforehand in a 35 × 1.1 mm precolumn packed with LiChrosorb RP-18 (d_p = 7 μm). 100 μl of water were introduced with 1) $5 \cdot 10^{-7}$%, 2) $2 \cdot 10^{-7}$%, and 3) no additional nitrobenzene. The detection was a) UV at 254 nm and b) electrochemical (E = −800 mV). (Reproduced from [165] with the permission of Elsevier Science Publishers.)

sample was then fed into the chromatograph system, which was based on a 250 × 1.1 mm column packed with LiChrosorb RP-18. A UV detector with wavelength 254 nm and an electrochemical detector with E = −800 mV both yielded concentrational limits of $2 \cdot 10^{-7}\%$. A chromatogram is given in Fig. 6.25, and it is clear that the nitrobenzene present is at the detection limit.

6.7.5. Analysis of Phthalic Acid Esters in River Water Using Off-Line Sample Concentration

The off-line system we described in Section 6.6 for investigating corticosteroids in blood serum [41] was applied to the analysis of 10-ml samples of river water for the esters of phthalic acid [40]. The sorbent chosen for the working column was TSK gel LS-III because, in contrast to silica gels, a polymer sorbent's adsorptive activity is not sensitive to the presence of water in the eluent. Figure 6.26 shows a chromatogram for a mixture of phthalic acid esters injected into the precolumn. Diethyl phthalate is not sorbed well because of its polarity; nevertheless, it could be detected at 1 ppb in a 10-ml sample, because it is retained quantitatively in the microbore column. A flow spectrophotometer with a measuring cell of 0.05-0.4 μl yielded a threshold sample mass of 25 and 40 ng for diethyl phthalate and dibutyl phthalate for a signal-to-noise ratio of 2. The quantities of ester could be determined from calibration curves drawn from test mixture data.

6.7.6. Antioxidants in Gasoline

Nakanishi [359] reported the use of microbore columns with sample concentration for measuring how much antioxidant is present in gasolines. The sample is concentrated in a 10 × 0.2 mm PTFE column packed with Develosil 60 (10 μm) and separated in a 150 × 0.26 mm fused-silica column packed with LiChrosorb RP-8 (5 μm). The separation of the antioxidants and metallic deactivator (four components) took 20 min. A 65:35:1 mixture of acetonitrile, water, and n-hexylamine was used at an elution rate of 2.1 μl/min. A spectrophotometric detector provided a threshold sample mass of 0.3 ng in 0.3 ml of gasoline (1 ppm). Nakanishi demonstrated that whether obtained from catalytic cracking or catalytic reforming, or the light fraction from straight-run gasoline, none of the gasolines contained antioxidants. The concentration of N-phenyl-N'-sec-butyl-p-phenylenediamine in gasoline was 8.6 μg/ml, and this value can be obtained in 10 min with an error of 2%.

6.7.7. Hydrocarbons in Coal Extracts

Hirose et al. [360] used 1800 × 0.2 mm fused-silica capillaries packed with Spherisorb C-18 (3 μm) to separate and identify the isomers of polycyclic hydrocarbons in coal extracts. A column efficiency of 225,000 plates made it possible to separate completely 64 out of 114 components within 11 h. A step gradient regime of acetonitrile in water at a rate of 1.1 μl/min was employed. Isomers containing 6 to 9 rings and with masses from 202 to 450 daltons were separated. The components were identified with a mass spectrometer, which nevertheless showed that many of the peaks contained mixtures of compounds.

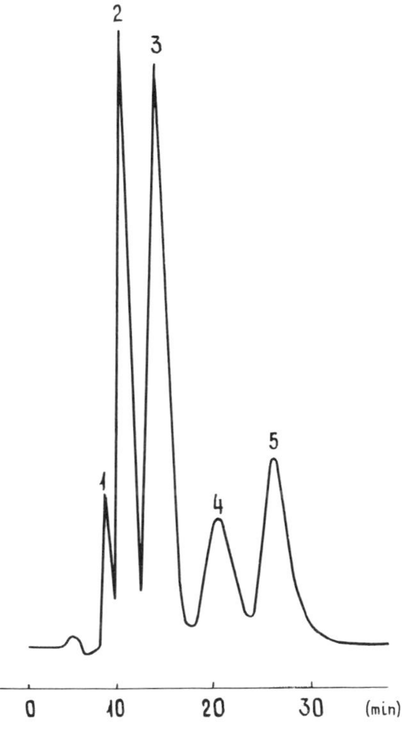

Fig. 6.26. Chromatogram of a standard mixture of phthalic acid esters taken on a 170 × 0.5 mm column packed with TSK gel LS-III. The eluent was a 5:1 methanol–dichloromethane system and was used at a flow rate of 6.25 μl/min. Sample concentrated beforehand on a 15 × 0.35 mm precolumn packed with Hitachi gel 3010. The 0.06-μl sample contained 1) 0.43% diethyl phthalate, 2) 2.8% dibutyl phthalate, 3) 5.8% dihexyl phthalate, 4) impurities in the dihexyl phthalate, and 5) 6.9% didecyl phthalate. (Reproduced from [362] with the permission of Elsevier Science Publishers.)

6.8. MICROBORE LIQUID CHROMATOGRAPHY APPLIED TO INORGANIC COMPOUNDS

This technique is useful for inorganic compounds too, because only a small quantity of eluent, usually corrosive, is consumed, and if neutron-activation analysis is being used, only small samples are needed. Moreover, because the technique is rapid, short-lived isotopes can be analyzed.

Ishii et al. have developed methods for separating halogen ions [43], alkali-metal ions [42], and rare-earth elements [44]. They used ion-exchange chromatography with both cationites and anionites on a sulfopolystyrene base. They packed Teflon columns, 0.5 mm ID, with sorbent using the slurry technique described in [38]. The sorbent was enclosed at top and bottom with quartz wool. A microsyringe, which delivered both the eluent and sample, was connected to the top of the column, while the bottom was drawn down to a 0.25 mm ID over a Bunsen burner flame. This served for the detector of beta radiation activity (when investigating radioactive isotopes obtained by neutron activation) or was connected to a color detection system (for rare-earth investigations). The radioactivity detector was described in Section 4.2.9. After passing through this detector the effluent was collected on a 30-mm-square piece of filter paper for gamma radiation analysis by a germanium semiconductor detector connected to an Amberra 4096-channel impulse-height analyzer.

6.8.1. Halide Ions

The radioactive elements to be used in neutron-activation analysis must be separated very quickly, especially when their half-lives are short. Thus, the radioactive halogens ^{38}Cl, ^{80}Br, and ^{128}I (whose half-lives are 37.2, 17.8, and 25 min, respectively) must be separated within 30 min of their formation.

A halogen separation on conventional liquid chromatographic columns takes 3 h in a $NaNO_3$ solution [361, 362], and 1 h if organic solvents are added to the solution [362]. Ishii et al. [43] developed an ion-exchange method in microbore columns for completely separating halide ions in less than 20 min.

The standards they used were a 3 M aqueous solution of NH_4Cl, a 10 M aqueous solution of NH_4Br, and a 10 M aqueous solution of NH_4I, ammonium thiosulfate being added to the last solution to keep it from being oxidized. A mixture containing the three ions was sealed into a PTFE tube 1 mm in diameter, and then placed in a TRIGA Mark-II reactor and irradiated by a stream of thermal neutrons (about $1.5 \cdot 10^{12}$ $cm^{-2} \cdot sec^{-1}$) for 5-25 min. They obtained the following:

Element	Mass no.	Half-life	Beta maximum, MeV
Chlorine	38	37.2 min	4.91
Bromine	80	17.8 min	2.00
Bromine	82	35.6 h	0.44
Iodine	128	25 min	2.12

A beta counter was used for ^{80}Br and a gamma spectrometric method for ^{82}Br.

The separation was carried out on a 59 × 0.5 mm Teflon column packed with TSK LS-222 anion-exchange sorbent (d_p = 6 μm), the eluent being a 1:1 mixture of 1.5 N $NaNO_3$ and acetone. Three halide ions were separated in 19 min. The separation coefficients obtained were Cl/Br = 1.3 and Br/I = 3.1. If only the iodine ion was to be determined, then a column only 12 mm long is needed. Using a 1:3 mixture of 1.5 N $NaNO_3$ and acetone as eluent, Ishii et al. obtained 0.08 μg of iodine in the presence of 34 μg of bromine and 41 μg of chlorine within 9.5 min. No other halide ions were observed in the gamma spectra of the separated ions.

6.8.2. Alkali Metals

Ishii et al. [42] developed a cation-exchange microbore chromatograph with a radioactivity detector to separate sodium, potassium, rubidium, and cesium.

To prepare radioactive solutions, crystals of sodium and potassium nitrates and rubidium and cesium sulfates were wrapped in aluminum foil and irradiated in an IRR-4 reactor by a flux of thermal neutrons ($1.8 \cdot 10^{13}$ neutrons/$cm^2 \cdot$sec) for 1 h. Finally, the crystals were dissolved in a small quantity of water. The following were obtained:

Element	Mass no.	Half-life	Beta maximum, MeV
Sodium	24	15 h	1.39
Potassium	42	12.5 h	3.52
Rubidium	86	18.7 h	1.78
Cesium	134	2.1 years	0.66

The isotopes were separated in a 72 × 0.5 mm column packed with Hitachi 2610 cationite. Because the eluent HCl is corrosive, it was not collected in the microsyringe, but in a 1300 × 0.5 mm ID PTFE tube, which was placed between the pump and the column. A special capillary measuring 15 × 0.31 mm ID (outside diameter 0.51 mm) was fitted to the end of the PTFE tube to take the sample. The pump and the first section of the tube containing the eluent were filled with benzene, and then the HCl was fed into the tube. The device provided an eluent rate of 8 μl/min. A 0.5-μl sample of the four metal

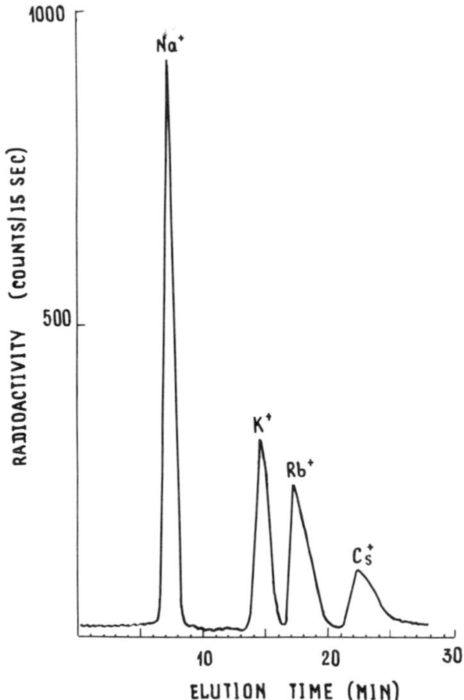

Fig. 6.27. Chromatogram of alkali metals separated using liquid chromatography on a 72 × 0.5 mm column packed with Hitachi N2610 cationite (d_p = 6 μm). Eluent: 0.7 N HCl. Flow rate: 8 μl/min. The sample was 0.1-2 μl of solution containing 0.07 μg Na, 0.2 μg K, 3.5 μg Rb, and 6.5 μg Cs. (Reproduced from [42] with the permission of Elsevier Science Publishers.)

ions was fed into the column, the free volume having been determined by the yield of ^{131}I. The metal ions were separated the best for an HCl eluent concentration of 0.7 N. A chromatogram of the four ions is shown in Fig. 6.27. It was obtained in 26 min with an elution rate of 8 μl/min, the separation coefficients being Na/K = 4.7, K/Rb = 1.2, and Rb/Cs = 1.4. If the elution rate is increased to 16 μl/min and an 80 × 0.5 mm column used, the ions can be separated in 15 min with the coefficients Na/K = 4.0, K/Rb = 1.0, and Rb/Cs = 1.3.

6.8.3. Rare-Earth Elements

Conventional ion-exchange techniques take more than 4 h to separate the rare-earth elements [363, 365, 366]. High-performance capillary chromatography can reduce this time to 80 min [364]. Ishii et al. [44] have shown that a promising method is to use microbore chromatography with radioactivity detection. They have separated the 16 elements on a cation-exchange microbore liquid chromatograph, using the cationite TSK LS-212, which is a sulfonated porous styrene–divinylbenzene matrix (d_p = 6 μm). Its capacity is a few milliequivalents per gram.

The radioactive isotopes were obtained by neutron activation by sealing several tens of microliters of a standard solution of ten elements (except Sc, Y, Ce, Pm, Gd, Tb, Tm) in a polyethylene tube 3 mm in diameter. The tube was irradiated either in an IRR-2 reactor for 20 min (thermal neutron flux $7 \cdot 10^{13}$ neutrons/cm^2·sec) or a TRIGH Mark-II reactor for 30 min (flux $1.2 \cdot 10^{12}$ neutrons/cm^2 × sec). A standard solution of Sc, Y, Gd, Tb, and Tm was placed in a silica vessel, carefully dried, sealed, and then irradiated in an IRR-4 reactor (flux $4 \cdot 10^{13}$ neutrons/cm^2·sec) for 5 h. The salts were then dissolved in water or dilute nitric acid. The ^{141}Ce and ^{147}Pm isotopes were obtained from the Japan Radioisotopic Association. All the isotopes, except for the ^{147}Pm, were mixed together and the resultant mixture diluted to a final overall concentration of 0.05-0.5 μCi/μl. A 0.22-μl sample of this solution was injected into the column. The sample thus contained ^{46}Sc, ^{90}Y, ^{140}La, ^{141}Ce, ^{142}Pr, ^{149}Nd, ^{147}Pm, ^{153}Sm, ^{152}Eu, ^{159}Gd, ^{160}Tb, ^{165}Dy, ^{166}Ho, ^{177}Er, ^{170}Tm, ^{175}Yb, ^{177}Yb, ^{176}Lu, and ^{177}Lu.

The isotopes were separated using gradient elution in a 0.4 M solution of α-hydroxyisobutyric acid (α-HIBA) with a programmed pH change from 3.1 to 6. The pH gradient was obtained by adding 0.4 M α-HIBA-NH_3, which has pH 6, from a 1-ml syringe into a 2-ml mixer filled with the 0.4 M α-HIBA, which has pH 2.2. The eluent was then transferred to the 1500 × 0.5 mm Teflon tube by a 250-μl pump. The tube was prepared by establishing a pH gradient in it. The tube was first filled with 0.4 M α-HIBA-NH_3 (pH 6). Then 0.4 M α-HIBA (pH 2.2) was added to the mixer, from which the pump moved it to the tube, until the pH reached 3.5. Then 40 μl of solution with pH 3.5 was fed into the tube. To make up the volume 40 μl of a prepared solution of 0.4 M α-HIBA-NH_3 with pH 3.5 was injected into the mixer with a micropipette. Using a 1-ml syringe 0.4 M α-HIBA (pH 2.2) was then added to the mixer until the pH reached 3.1. Finally, 40 μl of the pH 3.1 solution was fed into the tube.

In order to separate 14 of the lanthanides, a simpler gradient was needed. The column was filled with the pH 6 solution, and then the pH 2.2 solution was added to the mixer until the pH reached 3.2, using if need be the 1-ml syringe.

Ishii et al. showed that the best pHs for separating the elements in a 0.4 M solution of α-HIBA were 3.1 for Sc from Lu, 3.5 for Y from Dy, 4.0 for Sm from Eu, and 4.64 for Pr from Ce, while La could be quickly separated out at pH 6.0. Each of these separations took 4 min. The chromatogram in Fig. 6.28 shows the separa-

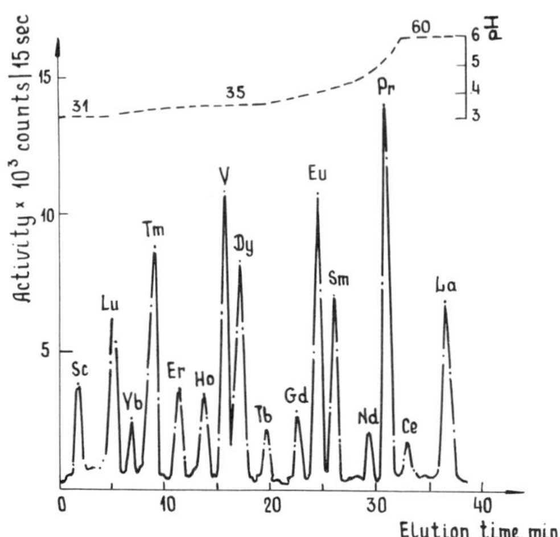

Fig. 6.28. Chromatogram of rare-earth separation using liquid chromatography on a 75 × 0.5 mm column packed with TSK LS-212 cationite (d_p = 6 μm). The eluent was a 0.4 M solution of α-HIBA in ammonia used in a gradient from pH 3.1 to 6.0 (dashed line) at a flow rate of 8 μl/min. The sample was 0.22 μl of solution containing 6-10 μg of rare earth. (Reproduced from [44] with the permission of Elsevier Science Publishers.)

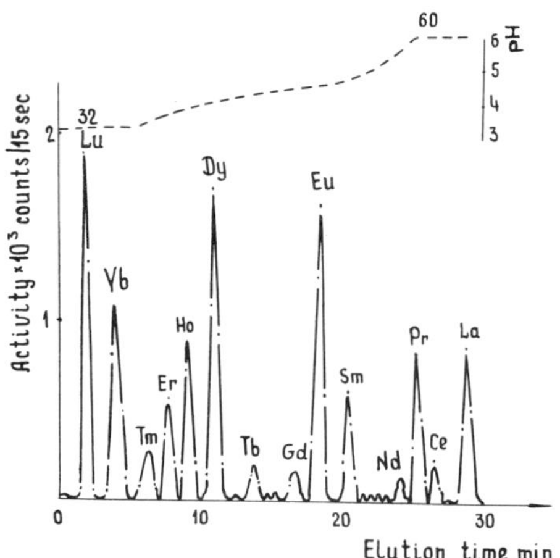

Fig. 6.29. Chromatogram for lanthanides taken on a 70 × 0.5 mm column packed with TSK LS-212 cationite (d_p = 6 μm). The eluent was a 0.4 N solution of α-HIBA in ammonia with gradient pH from 3.2 to 6.0 (dashed line). Other parameters as in Fig. 6.28. (Reproduced from [44] with the permission of Elsevier Science Publishers.)

tion of all 16 lanthanides and took 98 min. The separation coefficients were 1.8 for Lu/Yb, 1.0 for Y/Dy, 2.7 for Tb/Gd, and 1.9 for Pr/Ce. Figure 6.29 gives the chromatogram for the separation of the 14 lanthanides using the simpler gradient regime.

The radioactivity was detected in the same way as for the halide and alkali-metal ions. All the isotopes could be detected thus, except for ^{147}Pm, which is a very weak beta emitter (0.224 MeV) and so its radiation is absorbed by the walls of the detection chamber. However, the K_R values of this and the other elements indicate that the column could separate Pm, too. The gamma spectra of the effluent showed that all but the Yb peak, which contained some Dy, were pure. Several of these spectra were used to estimate quantitatively the amounts in the peaks.

Hirose et al. [226] have also reported using postcolumn colorimetry to detect the lanthanides using xylene orange. A separation of 15 lanthanides took 55 min.

The colored chelates were obtained by adding to the effluent a solution of 0.005 M xylene orange, 0.01 M cetylpyridinium bromide, and 20% methanol in a buffer solution made from equal volumes of 1.1 M NH_3 and 6 M NH_4Cl (pH 8.7). The colored chelates absorb wavelengths between 600 and 620 nm most strongly; however, the detection was carried out at 630 nm to reduce the background absorption. The detector signal was linear to 0.9 optical density units, and the threshold sample mass was 0.3-0.8 μg, being larger for the heavier elements (1.6 μg for Tb).

6.8.4. Metal Chloro Complexes

Using an ion-exchange chromatograph and an HCl eluent, to which $HClO_4$ was sometimes added, it was possible [367] to separate the chloro complexes of Fe, Ni, Cu, Sn, Sb, Hg, Pb, Co, and Bi on a special anionite made from the silica gel LiChrosorb Si-60 (d_p = 10 μm) by impregnating it with methyltrioctylammonium chloride. This sort of anionite is stable in the presence of concentrated mineral acids. Metal complexes with HCl absorb strongly between 210 and 220 nm. In order to protect the pump from corrosion by the acid, it was filled with water, which then pushed the gradated HCl (from 1 to 8 N) solution from a 3000 × 0.5 mm Teflon capillary into the packed column, 7-20 cm long and 0.5 mm internal diameter.

The connections between the Teflon capillaries and the tube for injecting the sample were made from platinum–iridium tubing, internal diameter 0.3 mm. The detector flow cell was made from a 10 × 0.3 mm silica tube connected to the end of the column. Note that the threshold sample mass in this analysis was 1000 times lower than for any known conventional liquid chromatographic analysis. We believe that the ion-exchange separation of chloro complexes may also be implemented on open tubular columns.

6.8.5. Nitrates in Water

The nitrate concentration in a 20-μl sample of water can be determined on a 150 × 2 mm column packed with Partisil-10 SAX with 0.001 M potassium phthalate (pH 3.95) as the eluent. The work by Cooke [368] is interesting because it demonstrates a new fluorometric detection system for species that do not absorb ultraviolet light (e.g., NO_2^-, NO_3^-, and Cl^-). The effluent is detected by the increase in UV transparency of the eluent when the sodium phthalate is sorbed onto the anionite and the NO_2^-, NO_3^-, or Cl^- ions released. The detection wavelength used was 265 nm, the detection cell volume 8 μl, and the resultant threshold concentration 0.1 ppm.

6.8.6. Ultramicrobore Analyses of Inorganic Ions Using a UV Detector

Rokushika et al. [205] have used a 280 × 0.19 mm fused-silica column packed with the strong anionite resin YEW AX-1 (25 meq/g and d_p = 10 μm) to separate NO_2^-, Br^-, NO_3^-, and SO_4^{2-}. A 0.004 M solution of sodium carbonate and sodium bicarbonate was used at an elution rate of 4 μl/min, with the sample volume being 0.05 μl. A 0.2-mm silica capillary was used as the detection cell, the sensitivity being greatest for a wavelength of 195 nm. A separation of NO_2^-, Br^-, NO_3^-, and SO_4^{2-} required 4 min, for which the threshold concentrations were 10 ppm for the NO_2^-, Br^-, and NO_3^-, and 150 ppm for the SO_4^{2-}.

In terms of the NO_3^- the threshold sample mass obtained in this paper is 1/5 that obtained in Cooke's paper [368] for a sample volume 1/3 of that used by Cooke. If the sample volume used by Rokushika et al. is increased to the amount used by Cooke, then the threshold mass would be 25 times less than that Cooke achieved using a 2-mm column. Moreover, this result was achieved using a UV detector whose optical path was 50-fold shorter than Cooke's, though some of this better efficiency is apparently because the NO_2^- ion was detected directly at a wavelength of 195 nm.

Chapter 7

CONCLUSION

In order to determine the most promising areas of application of capillary liquid chromatography, we shall summarize the pros and cons of microbore columns.

An indubitable advantage is the economy of eluent and stationary phase, which allows us to decrease by tenfold the consumption of solvent, reduce the toxic and fire dangers, and employ exotic and expensive solvents that can improve the selectivity or detection capacities of the system. Another distinct advantage is the simplicity of connecting a microbore liquid chromatograph to a mass spectrometer via a direct liquid introduction interface, to a Fourier-transform IR spectrometer, or to a gas-chromatography flame detector. Temperature programming is also easier with capillary columns because of the efficient radial heat transfer. When a multicomponent mixture is being separated, this method is often a good alternative to gradient elution.

Capillary liquid chromatography is more sensitive than the conventional high-performance variant, but only if a concentration-sensitive detector is used and a constant sample mass is introduced in the maximum permissible volume. The anticipated greater sensitivity would be even more so if the concentrational sensitivity of the detector is maintained when its flow cell is miniaturized. For spectrophotometric detectors this means that the optical path length must be kept close to 1 cm, but this can only be achieved if the optical scheme and the design of the detector cell are ingenious (e.g., ISCO's μLC-10).

On the other hand, for trace analyses (a constant concentration sample is introduced) capillary liquid chromatography with concentrational detectors is not more sensitive than the conventional high-performance variant, and is decidedly worse if mass-sensitive detectors or fluorometers are employed.

Fast chromatography has been successfully implemented on both conventional packed and microbore columns. It may be that the better radial heat-exchange of the microbore columns will mean that the productivity (efficiency per unit time) of the chromatographic process will be somewhat better in them than in conventional columns. However, this has not yet been convincingly demonstrated experimentally.

Nevertheless, microbore chromatography is more convenient for fast analyses because of the considerably lower solvent consumption and because it is unnecessary to have special pumps with deliveries in excess of 10 ml/min.

As far as superefficient systems are concerned, open tubular capillary columns (d_c = 10 μm) are clearly better for efficiencies in excess of 500,000 theoretical plates. Even though systems for sample introduction and solute detection have been developed for open tubular capillary columns, this area remains one for the "ace" experimenters. This is because of the difficulties involved in the preparation of the columns and the tribulations that arise because they can clog up. Therefore, if no radically new way of overcoming these volumetric limitations of open tubular capillaries is found, then it seems unlikely that this method will become widely used.

By contrast, packed columns look very promising for superefficient liquid chromatography. Although columns with the "conventional" diameter have been connected together into successful systems and described in the literature, superefficient microbore systems are very economical. Moreover, considerably longer individual sections of ultramicrobore fused-silica columns (up to 200 cm) have been well packed with microparticle sorbents than have conventional columns (50 cm), and hence the number of concatenations is less. Finally, the relative ease with which a superefficient microbore system can be connected to a mass spectrometer or a Fourier-transform IR detector is particularly valuable, since these systems can be used to resolve and identify complicated multicomponent mixtures of compounds that are chemically very close.

Thus, microbore chromatographs are useful for the following analytical or purely chromatographic tasks:

1) developing new analysis techniques with rare sorbents or eluents;

2) testing new types of sorbent;

3) investigating the regimes for packing columns;

4) separating complicated mixtures and identifying them from their mass or IR spectra;

5) doing ultrasensitive analyses with concentration-sensitive detectors, given a sample with a certain mass;

6) doing very sensitive selective analyses of compounds containing sulfur, phosphorus, nitrogen, or halogens using gas-chromatography detectors;

7) doing routine analyses with large numbers of samples, particularly when rare solvents are being used;

8) doing fast analyses, such as drug monitoring, or for kinetically controlling a chemical reaction;

9) doing analyses with toxic, radioactive, or inflammable compounds, one of which might be the solvent; and

10) using temperature programming.

Microbore liquid chromatography does not have any particular advantage, other than solvent economy, for trace analyses unless the sample is enriched. It may even be less sensitive than conventional systems if a flow-sensitive detector is used.

In order for the advantages of microbore technology to be widely applied it is essential that the ancillary equipment for it be improved, i.e., the concentrational and mass sensitivities of the detectors need to be increased, precise gradient solvent delivery systems created, the microbore chromatographs automated, and rapid systems for processing the data developed.

The instrument suppliers that are still trying to unify their equipment for microbore, conventional, and even preparative high-performance liquid chromatographies are hindering the advance of microbore chromatography.

Progress can be seen in the creation of detector systems (ISCO), syringe pumps (Brownlee Labs, ISCO), and injectors for microbore columns. Fascinating possibilities are opening up with the use of new detectors, such as the parallel-opposed electrode electrochemical detector, mass detectors (evaporation analyzers), and laser detectors.

Meanwhile, there is considerable scope for enhancing the performance of microbore columns. The efficiencies of packed microbore columns might be raised two- or threefold if the packing procedure could be optimized.

For open tubular capillary columns, the situation is at an impasse because it is very difficult to combine the very rigid theoretical requirements on the size of their diameters and convenience and reliability of use. It is not impossible, however, that some new ideas may arise for getting around this paradox.

The supercritical-fluid technique looks very promising when applied to capillary columns. Increasing the diffusivity of the solute and reducing the viscosity of the eluent decrease the time needed for the analysis without requiring an increase in pressure. The optimum diameter of open tubular columns using this technique turns out to be several times greater than that when using liquid chromatography, thus facilitating their use.

Thus, our prognosis for the development of capillary liquid chromatography is very optimistic. And on that note it is nice to end our work on this book.

REFERENCES

1. P. Kucera (ed.), Microcolumn High-Performance Liquid Chromatography, Journal of Chromatography Library, 28, Elsevier, Amsterdam (1984).
2. R. P. W. Scott (ed.), Small-Bore Liquid Chromatography Columns: Their Properties and Uses, Chem. Anal. Series, Vol. 42, Wiley, New York (1984).
3. F. J. Yang, "Microbore column HPLC," High-Resol. Chromatogr., Chromatogr. Commun., 6, 348-358 (1983).
4. B. G. Belen'kii (Belenky), É. S. Gankina, O. I. Kurenbin, and V. G. Mal'tsev, "Capillary liquid chromatography, problems and prospects," in: Chromatography, the State of the Art, H. Kalacz and L. S. Ettre (eds.), Publishing House of the Hungarian Academy of Sciences, Budapest (1984), pp. 841-885.
5. R. P. W. Scott, "Small-bore columns in liquid chromatography," Adv. Chromatogr. (N.Y.), 22, 247-294 (1983).
6. B. G. Belen'kii, É. S. Gankina, and V. G. Mal'tsev, "Capillary liquid chromatography," Zh. Vses. Khim. Ova., 28, 43-47 (1983).
7. S. V. Kuz'min, V. V. Matveev, E. K. Pressman, and L. S. Sandakhchiev, "Simple procedure of quantitative chromatographic ultramicroanalysis," Biokhimiya, 34, 706-711 (1969).
8. L. S. Sandakhchiev, in: Ultramicroanalysis of Nucleic Acids, D. G. Knorre and T. V. Venkstern (eds.), Nauka, Moscow (1973), pp. 77-94.
9. M. A. Grachev, in: Ultramicroanalysis of Nucleic Acids, D. G. Knorre and T. V. Venkstern (eds.), Nauka, Moscow (1973), pp. 104-122.
10. S. V. Kuz'min, in: Ultramicroanalysis of Nucleic Acids, D. G. Knorre and T. V. Venkstern (eds.), Nauka, Moscow (1973), pp. 95-103.

11. M. L. Aleksandrov and V. A. Pavlenko, in: Instruments for Scientific Investigations, V. A. Pavlenko (ed.), Materials of International Scientific and Technical Conference of the Member Countries of the Council for Mutual Economic Aid on Scientific Instruments – Nauchpribor-78, SÉV, Moscow (1980), pp. 399-404.
12. V. G. Mal'tsev, E. M. Koroleva, B. G. Belen'kii, R. G. Vinogradova, and M. B. Ganitskii, "Ultramicroanalysis of DNS amino acids by microcolumn hydrophobic chromatography with analytical sensitivity of 10^{-13} mole," Bioorg. Khim., 9, 186-195 (1983).
13. J. J. Kever, É. S. Gankina, and B. G. Belen'kii, "Microcolumn exclusion chromatography of polymers," Vysokomol. Soedin., A23, 234-236 (1961).
14. P. Kucera, "Design and use of short microbore columns in liquid chromatography," J. Chromatogr., 198, 93-109 (1980).
15. R. P. W. Scott, "Microbore columns in liquid chromatography," J. Chromatogr. Sci., 18, 49-54 (1980).
16. P. Kucera and G. Manius, "High-resolution reversed-phase liquid chromatography utilizing microbore column concatenation," J. Chromatogr., 216, 9-21 (1981).
17. F. J. Yang, "Fused-silica narrow-bore microparticle-packed column high-performance liquid chromatography," J. Chromatogr., 236, 265-277 (1982).
18. T. Takeuchi and D. Ishii, "High-performance micro-packed flexible columns in liquid chromatography," J. Chromatogr., 213, 25-33 (1981).
19. J. C. Gluckman, A. Hirose, V. L. McGuffin, and M. Novotny, "Performance evaluation of slurry-packed capillary columns for liquid chromatography," Chromatographia, 17, 303-309 (1983).
20. T. Tsuda and G. Nakagawa, "Open-tubular liquid chromatography with 5-10 μm ID columns," J. Chromatogr., 268, 369-374 (1983).
21. R. Tijssen, J. P. A. Bleumer, A. L. C. Smit, and M. E. Van Kreveld, "Microcapillary liquid chromatography in open tubular columns with diameters of 10-50 μm. Potential application to chemical ionization mass-spectrometric detection," J. Chromatogr., 218, 137-167 (1981).
22. J. H. Knox and M. T. Gilbert, "Kinetic optimization of straight open-tubular liquid chromatography," J. Chromatogr., 186, 405-418 (1979).
23. J. H. Knox, "Theoretical aspects of LC with packed and open small-bore columns," J. Chromatogr. Sci., 18, 453-461 (1980).
24. T. Tsuda and M. Novotny, "Packed microcapillary columns in high-performance liquid chromatography," Anal. Chem., 50, 271-275 (1978).
25. Y. Hirata and M. Novotny, "Techniques of capillary liquid chromatography," J. Chromatogr., 186, 521-528 (1979).
26. Y. Hirata, M. Novotny, T. Tsuda, and D. Ishii, "Packed microcapillary columns with different selectivities for liquid chromatography," Anal. Chem., 51, 1807-1809 (1979).
27. M. Novotny, "Capillary HPLC: Columns and related instrumentation," J. Chromatogr. Sci., 18, 473-478 (1980).

28. V. L. McGuffin and M. Novotny, "Optimization and evaluation of packed capillary columns for high-performance liquid chromatography," J. Chromatogr., 255, 381-393 (1983).
29. G. Guiochon, "Conventional packed columns vs. packed or open-tubular microcolumns in liquid chromatography," Anal. Chem., 53, 1318-1325 (1981).
30. G. Guiochon, "Preparation and operation of liquid chromatographic columns of very high efficiency," J. Chromatogr., 185, 3-26 (1979).
31. G. Guiochon and H. Colin, in: Microcolumn High-Performance Liquid Chromatography, P. Kucera (ed.), Journal of Chromatography Library, 28, Elsevier, Amsterdam (1984), pp. 1-38.
32. H. Stegemann and G. Bernhard, "The separation and analysis of amino acids in ultramicroscale," Microchim. Acta, 555-563 (1961).
33. E. Kirsten and R. Kirsten, "Analytical chromatography of amino acids of the order of magnitude of nanomoles," Biochem. Z., 339, 287-304 (1964).
34. C. G. Horvath, B. A. Preiss, and S. R. Lipsky, "Fast liquid chromatography; an investigation of operating parameters and the separation of nucleotides on pellicular ion exchangers," Anal. Chem., 39, 1422-1428 (1967).
35. G. Nota, G. Marrino, V. Buonocore, and A. Ballio, "Liquid—solid chromatography with open glass capillary columns. Separation of 1-dimethylamino-5-naphthylsulfonylamino acids," J. Chromatogr., 46, 103-106 (1970).
36. S. V. Kuzmin, "Double-beam spectrophotometer," US Patent No. 3749497 (1973).
37. D. Ishii, "Simple miniaturization of high-performance liquid chromatography," Application notes, No. 9, 1-9 (1976).
38. D. Ishii, K. Asai, K. Hibi, T. Yonokuchi, and M. Nagaya, "A study of micro-high-performance liquid chromatography. I. Development of technique for miniaturization of high-performance liquid chromatograph," J. Chromatogr., 144, 157-168 (1977).
39. D. Ishii, K. Hibi, K. Asai, and T. Yonokuchi, "Studies of micro-high-performance liquid chromatography. II. Applications to gel-permeation chromatography of techniques developed for micro-high-performance liquid chromatography," J. Chromatogr., 152, 147-154 (1978).
40. D. Ishii, K. Hibi, K. Asai, and M. Nagaya, "Studies of micro-high-performance liquid chromatography. III. Development of a 'micro-precolumn' method for pretreatment of samples," J. Chromatogr., 152, 341-348 (1978).
41. D. Ishii, K. Hibi, K. Asai, and M. Nagaya, "Studies of micro-high-performance liquid chromatography. IV. Application of the micro-precolumn method to the analysis of corticosteroids in serum," J. Chromatogr., 156, 173-180 (1978).
42. D. Ishii, A. Hirose, K. Hibi, and Y. Iwasaki, "Studies of micro-high-performance liquid chromatography. V. Design of a micro-scale liquid chromatograph and its application to cation-exchange separation of alkali metals," J. Chromatogr., 157, 43-50 (1978).

43. D. Ishii, A. Hirose, and J. Horiuchi, "Application of microscale liquid chromatographic technique to the anion-exchange separation of halide ions," J. Radioanal. Chem., 45, 7-14 (1978).
44. D. Ishii, A. Hirose, and Y. Iwasaki, "Cation-exchange separation of 16 rare-earth metals by microscale high-performance liquid chromatography," J. Radioanal. Chem., 46, 41-49 (1978).
45. T. Takeuchi and D. Ishii, "Ultra-micro-high-performance liquid chromatography," J. Chromatogr., 190, 150-155 (1980).
46. T. Takeuchi and D. Ishii, "Micro-HPLC with long micro-packed flexible fused-silica columns," J. Chromatogr., 238, 409-413 (1982).
47. D. Ishii and T. Takeuchi, "High-performance liquid chromatography using flexible fused-silica micro-packed columns," J. Chromatogr., 255, 349-358 (1983).
48. T. Takeuchi and D. Ishii, "A study on fast micro-high-performance liquid chromatography," High-Resol. Chromatogr., Chromatogr. Commun., 6, 683-685 (1983).
49. T. Takeuchi and D. Ishii, "Application of ultra-micro-high-performance liquid chromatography to trace analysis," J. Chromatogr., 218, 199-209 (1981).
50. T. Takeuchi and D. Ishii, "Gradient separations of complex mixtures by micro high-performance liquid chromatography," High-Resol. Chromatogr., Chromatogr. Commun., 6, 310-315 (1983).
51. Y. Hirata and E. Sumiya, "Temperature-programmed reversed-phase liquid chromatography with packed fused-silica columns," J. Chromatogr., 267, 125-131 (1983).
52. R. P. W. Scott and P. Kucera, "Mode of operation and performance characteristics of microbore columns for use in liquid chromatography," J. Chromatogr., 169, 51-72 (1979).
53. R. P. W. Scott and P. Kucera, "Use of microbore columns for the separation of substances of biological origin," J. Chromatogr., 185, 27-41 (1979).
54. R. P. W. Scott, P. Kucera, and M. Munroe, "Use of microbore columns for rapid liquid chromatographic separations," J. Chromatogr., 186, 475-487 (1979).
55. I. Halász and G. Maldener, "Packing method for coupled macrobore liquid chromatography columns," Anal. Chem., 55, 1842-1847 (1983).
56. J. C. Kraek, H. Poppe, and F. Smedes, "Construction of columns with very large plate numbers. Theory and practice," J. Chromatogr., 122, 147-159 (1976).
57. I. Halász, R. Endele, and J. Asshauer, "Ultimate limits in high-pressure liquid chromatography," J. Chromatogr., 112, 37-60 (1975).
58. R. A. Hartwick and D. D. Dezaro, in: Microcolumn High-Performance Liquid Chromatography, P. Kucera (ed.), Journal of Chromatography Library, 28, Elsevier, Amsterdam (1984), pp. 75-110.
59. D. D. Dezaro, C. Hora, and R. A. Hartwick, "Kinetically optimized microbore HPLC in the high-speed monitoring of theophylline," Clin. Chem., 29, 1158-1162 (1983).

60. K. Jinno and M. Ishigaki, "Direct monitoring of reaction processes with micro-HPLC technique," J. Liq. Chromatogr., 6, 1065-1073 (1983).
61. C. E. Reese and R. P. W. Scott, "Microbore columns – design, construction, and operation," J. Chromatogr. Sci., 18, 479-486 (1980).
62. J. Bowermaster and H. M. McNair, "Temperature-programmed microbore HPLC. Part I," J. Chromatogr. Sci., 22, 165-171 (1984).
63. R. Roumeliotis, M. Chatziathanassion, and K. K. Unger, "Preparation and evaluation of assembled packed microbore columns in HPLC," Chromatographia, 19, 145-151 (1984).
64. J. Bowermaster and H. M. McNair, "Microbore high-performance liquid chromatographic columns: Speed, efficiency, sensitivity, and temperature programming," J. Chromatogr., 279, 431-438 (1983).
65. C. Eckers, K. K. Cuddy, and J. D. Henion, "Practical microbore column HPLC: System development and drug applications," J. Liq. Chromatogr., 6, 2383-2409 (1983).
66. H. Menet, P. Gareil, M. Caude, and R. Rosset, "Packing and performance of microbore columns for adsorption and partition chromatography," Chromatographia, 18, 73-80 (1984).
67. H. E. Schwartz, B. L. Karger, and P. Kucera, "Gradient elution chromatography with microbore columns," Anal. Chem., 55, 1752-1760 (1983).
68. J. D. Henion and T. Wachs, "Micro liquid chromatography/mass spectrometry diaphragm probe interface," Anal. Chem., 53, 1963-1965 (1981).
69. W. M. Niessen and H. Poppe, "Open-tubular liquid chromatography–mass spectrometry using direct liquid introduction," J. Chromatogr., 323, 37-47 (1985).
70. M. Krejči, K. Tesařik, M. Rusek, and J. Pajurek, "Flow characteristics and technology of capillary columns with inner diameters less than 15 μm in liquid chromatography," J. Chromatogr., 218, 167-179 (1981).
71. V. L. McGuffin and M. Novotny, "Micro-column high-performance liquid chromatography and flame-based detection principles," J. Chromatogr., 218, 179-189 (1981).
72. V. L. McGuffin and M. Novotny, "Flame emission detection in microcolumn liquid chromatography," Anal. Chem., 53, 946-951 (1981).
73. R. Tijssen, "Liquid chromatography in helically coiled open tubular columns," Sep. Sci. Technol., 13, 681-685 (1978).
74. K. Hibi, D. Ishii, I. Fujishima, T. Takeuchi, and T. Nakanishi, "Studies of open-tubular micro-capillary liquid chromatography. I. The development of open-tubular micro-capillary liquid chromatography," High-Resol. Chromatogr., Chromatogr. Commun., 1, 21-27 (1978).
75. T. Tsuda, K. Hibi, T. Nakanishi, T. Takeuchi, and D. Ishii, "Studies of open-tubular micro-capillary liquid chromatography. II. Chemically bonded octadecylsilane stationary phase," J. Chromatogr., 158, 227-232 (1978).

76. K. Hibi, T. Tsuda, T. Takeuchi, T. Nakanishi, and D. Ishii, "Studies of open-tubular micro-capillary liquid chromatography. III. β,β'-Oxydipropionitrile and ethylene glycol stationary phases," J. Chromatogr., 175, 105-111 (1979).
77. D. Ishii, T. Tsuda, K. Hibi, T. Takeuchi, and T. Nakanishi, "Studies of open-tubular micro-capillary liquid chromatography," High-Resol. Chromatogr., Chromatogr. Commun., 2, 371-376 (1979).
78. D. Ishii, T. Tsuda, and T. Takeuchi, "Studies of open-tubular micro-capillary liquid chromatography. IV. Soda—lime glass columns treated with alkaline solution," J. Chromatogr., 185, 73-78 (1979).
79. T. Takeuchi, M. Kumaki, and D. Ishii, "Role of column temperature in open-tubular microcapillary LC," J. Chromatogr., 235, 309-313 (1982).
80. K. Hibi, D. Ishii, and T. Tsuda, "Alumina and support-coated open-tubular columns in open-tubular micro-capillary liquid chromatography," J. Chromatogr., 189, 175-185 (1980).
81. T. Takeuchi, K. Matsuoka, Y. Watanabe, and D. Ishii, "Studies in open-tubular micro-capillary liquid chromatography. VI. Styrene—divinylbenzene copolymer stationary phase," J. Chromatogr., 192, 127-134 (1980).
82. D. Ishii, T. Tsuda, K. Hibi, T. Takeuchi, and T. Nakanishi, in: Glass Capillary Chromatography, R. Kaiser (ed.), Institute of Chromatography, Bad Dürkheim (1979), pp. 161-181.
83. T. Tsuda, K. Tsuboi, and G. Nakagawa, "Open-tubular micro-capillary liquid chromatography with 20 μm ID columns," J. Chromatogr., 214, 283-290 (1981).
84. D. Ishii and T. Takeuchi, "Open-tubular capillary LC," J. Chromatogr. Sci., 18, 462-472 (1980).
85. T. Takeuchi and D. Ishii, "Chemically bonded octadecylsilane and polyimine stationary phases for open-tubular microcapillary liquid chromatography," J. Chromatogr., 279, 439-448 (1983).
86. T. Takeuchi and D. Ishii, "Open-tubular dynamically modified silica columns in liquid chromatography," High-Resol. Chromatogr., Chromatogr. Commun., 6, 631-632 (1983).
87. T. Tsuda and G.Nakagawa, "Open-tubular micro-capillary liquid chromatography with 30-40 μm ID columns," J. Chromatogr., 199, 249-258 (1980).
88. T. Takeuchi, D. Ishii, and T. Nakanishi, "Dynamically modified silica columns in open-tubular capillary liquid chromatography," J. Chromatogr., 281, 73-81 (1983).
89. T. Takeuchi, H. Kitamura, T. Spitzer, and D. Ishii, "Use of cross-linked polysiloxane stationary phases in open-tubular capillary liquid chromatography," High-Resol. Chromatogr., Chromatogr. Commun., 6, 666-668 (1983).
90. D. Ishii and T. Takeuchi, "Study of the performance of cation exchange columns in open-tubular micro-capillary liquid chromatography," J. Chromatogr., 218, 183-199 (1981).
91. M. Krejči, K. Tesařik, and J. Pajurek, "Open-tubular columns in liquid chromatography," J. Chromatogr., 191, 17-23 (1980).

92. K. Tesařik, "Preparation of glass capillary columns for liquid chromatography," J. Chromatogr., 191, 25-30 (1980).
93. K. Slaiš and M. Krejči, "Electrochemical cell with effective volume less than 1 nl for liquid chromatography," J. Chromatogr., 235, 21-29 (1982).
94. J. W. Jorgenson and E. J. Guthrie, "Liquid chromatography in open-tubular columns, theory of column optimization with limited pressure and analysis time, and fabrication of chemically bonded reversed-phase columns on etched borosilicate glass capillaries," J. Chromatogr., 255, 335-348 (1983).
95. J. C. Giddings, J. P. Chang, M. N. Myers, J. M. Davis, and K. D. Caldwell, "Capillary liquid chromatography in field flow fractionation-type channels," J. Chromatogr., 255, 359-379 (1983).
96. R. Tijssen, J. P. A. Bleumer, and M. E. Van Kreveld, "Separation by flow (hydrodynamic chromatography) of macromolecules performed in open microcapillary tubes," J. Chromatogr., 260, 297-304 (1983).
97. R. F. Meyer, P. B. Champlin, and R. A. Hartwick, "Theory of multicapillary columns for HPLC," J. Chromatogr. Sci., 21, 433-438 (1983).
98. F. J. Yang, "On-column detection using a fused-silica column," High-Resol. Chromatogr., Chromatogr. Commun., 4, 83-85 (1981).
99. F. J. Yang, "Method and apparatus for on-column detection in liquid chromatography," US Patent No. 4375163 (1983).
100. E. J. Guthrie and J. W. Jorgenson, "On-column fluorescence detector for open-tubular capillary liquid chromatography," Anal. Chem., 56, 483-486 (1984).
101. E. J. Guthrie, J. W. Jorgenson, and P. R. Dluzneski, "On-column helium-cadmium laser fluorescence detector for open-tubular capillary liquid chromatography," J. Chromatogr. Sci., 22, 171-176 (1984).
102. L. A. Knecht, E. J. Guthrie, and J. W. Jorgenson, "On-column electrochemical detector with a single graphite fiber electrode for open-tubular liquid chromatography," Anal. Chem., 56, 479-482 (1984).
103. A. Manz and W. Simon, "Picoliter cell volume potentiometric detector for open-tubular column LC," J. Chromatogr. Sci., 21, 326-330 (1983).
104. Z. Fröbe, K. Richon, and W. Simon, "Anion-selective microelectrodes as femtoliter cell volume detectors for open-tubular column LC," Chromatographia, 17, 467-468 (1983).
105. Sj. van der Wal and F. Yang, "Gradient elution system for capillary and micro HPLC," High-Resol. Chromatogr., Chromatogr. Commun., 6, 216-217 (1983).
106. Y. Hirata and F. Nakata, "Gradient elution system for high-performance liquid chromatography with narrow-bore packed columns," J. Chromatogr., 294, 357-361 (1984).

107. T. Takeuchi and D. Ishii, "Continuous gradient elution in micro high-performance liquid chromatography," J. Chromatogr., 253, 41-47 (1982).

108. V. L. McGuffin and M. Novotny, "Nanoliter injection system for micro-column liquid chromatography," Anal. Chem., 55, 580-583 (1983).

109. T. Takeuchi, Y. Hirata, and Y. Okimura, "On-line coupling of a micro-liquid chromatograph and mass spectrometer through a jet separator," Anal. Chem., 50, 659-660 (1978).

110. S. Tsuge, Y. Hirata, and T. Takeuchi, "Vacuum nebulizing interface for direct coupling of micro-liquid chromatograph and mass spectrometer," Anal. Chem., 51, 166-169 (1979).

111. K. H. Schaffer and K. Levsen, "Direct coupling of a micro high-performance liquid chromatograph and a mass spectrometer," J. Chromatogr., 206, 245-252 (1981).

112. Yu. Yoshida, H. Yoshida, S. Tsuge, T. Takeuchi, and K. Mochizuku, "Direct measurement of mass fragmentograms for eluents from a micro-liquid chromatograph using an improved nebulizing interface," High-Resol. Chromatogr., Chromatogr. Commun., 3, 16-20 (1980).

113. T. Takeuchi, D. Ishii, A. Saito, and T. Ohki, "Direct coupling of an ultra-micro-high-performance liquid chromatograph and a mass spectrometer," High-Resol. Chromatogr., Chromatogr. Commun., 5, 91-92 (1982).

114. F. Heresch and J. F. K. Huber, in: Recent Developments in Mass Spectrometry in Biochemistry, Medicine, and Environmental Research, A. Frigerio (ed.), Elsevier, Amsterdam (1983), pp. 185-198.

115. A. P. Bruins and B. F. H. Drenth, "Experiments with the combination of a JASCO micro-liquid chromatograph and a quadrupole mass spectrometer," Int. J. Mass Spectrom. Ion Phys., 46, 213-216 (1983).

116. A. P. Bruins and B. F. H. Drenth, "Experiments with the combination of a micro liquid chromatograph and a chemical ionization quadrupole mass spectrometer using a capillary interface for direct liquid introduction. Some theoretical considerations concerning the evaporation of liquids from capillaries into vacuum," J. Chromatogr., 271, 71-82 (1983).

117. M. S. Lant, D. E. Games, S. A. Westwood, and B. J. Woodhall, "Microbore high-performance liquid chromatography—mass spectrometry," Int. J. Mass Spectrom. Ion Phys., 46, 189-192 (1983).

118. K. Levsen and K. H. Schaefer, "Direct coupling of a micro high-performance liquid chromatograph and a mass spectrometer," Int. J. Mass. Spectrom. Ion Phys., 46, 209-212 (1983).

119. F. R. Sugnaux, D. S. Skrabalak, and J. D. Henion, "Direct liquid introduction micro-liquid chromatography—mass spectrometry coupling. Optimization of droplet desolvation and instrumental parameters for high sensitivity," J. Chromatogr., 264, 357-376 (1983).

120. N. J. Alcock, L. Corbelli, D. E. Games, M. S. Lant, and S. A. Westwood, "Liquid chromatography/mass spectrometry using glass-lined stainless steel microbore columns," Biomed. Mass Spectrom., 9, 499-504 (1982).
121. P. Hirter, H. J. Walther, and D. Dätwyler, "Micro-column liquid chromatography—mass spectrometry using a capillary interface," J. Chromatogr., 323, 89-98 (1985).
122. H. Alborn and G. Stenhagen, "Direct coupling of packed fused-silica liquid chromatographic columns to a magnetic sector mass spectrometer and application to polar thermolabile compounds," J. Chromatogr., 323, 47-66 (1985).
123. R. Tiebach, W. Blass, and M. Kellert, "Design and construction of an interface for direct liquid introduction coupling of a microbore high-performance liquid chromatograph to a quadrupole mass spectrometer," J. Chromatogr., 323, 121-126 (1985).
124. N. Teramae and S. Tanaka, "On-line coupling of a micro-liquid chromatograph to a Fourier-transform infrared spectrophotometer," Spectrosc. Lett., 13, 117-125 (1980).
125. K. Jinno and C. Fujimoto, "Infrared monitoring in microcapillary liquid chromatography," Chromatographia, 17, 259-261 (1983).
126. R. S. Brown and L. T. Taylor, "Microbore liquid chromatography with flow cell Fourier transform infrared spectrometric detection," Anal. Chem., 55, 1492-1497 (1983).
127. K. Jinno, C. Fujimoto, and G. Uematsu, "Micro-HPLC/FTIR," Int. Lab., 14, No. 7 (March), 48-55 (1984).
128. C. C. Johnson and L. T. Taylor, "Normal phase liquid chromatography/Fourier transform infrared spectrometry for analysis of nonpolar material with semipreparative, analytical, and microbore columns," Anal. Chem., 55, 436-441 (1983).
129. S. Folestad, L. Johnson, and B. Josefsson, "Laser induced fluorescence detection for conventional and micro-column liquid chromatography," Anal. Chem., 54, 925-928 (1982).
130. S. Folestad, L. Johnson, B. Josefsson, and B. Galle, in: Proceedings of the Fourth International Symposium on Capillary Chromatography, R. E. Kaiser (ed.), Institute of Chromatography, Bad Dürkheim (1981), pp. 405-427.
131. L. W. Hershberger, J. B. Gallis, and G. D. Christian, "Submicroliter flow-through cuvette for fluorescence monitoring of high-performance liquid chromatographic effluents," Anal. Chem., 51, 1444-1446 (1979).
132. M. Goto, G. Zou, and D. Ishii, "Current amplification in dual electrochemical detector for micro high-performance liquid chromatography," J. Chromatogr., 268, 157-167 (1983).
133. E. M. Koroleva, V. G. Mal'tsev, and B. G. Belen'kii (Belenky), "Micro-chromatographic analysis of DNS amino acids with a sensitivity of 10^{-13} mole," J. Chromatogr., 242, 145-152 (1982).
134. J. J. Kever, B. G. Belen'kii (Belenky), É. S. Gankina, L. Z. Vilenchik, O. I. Kurenbin, and T. P. Zhmakina, "Determination of molecular-weight distribution of polymers by microcolumn exclusion chromatography," J. Chromatogr., 207, 145-147 (1981).

135. J. J. Kever, B. G. Belen'kii (Belenky), E. S. Gankina, L. Z. Vilenchik, O. G. Kurenbin, and T. P. Zhmakina, "Determination of the molecular-weight distribution of polymers by high-performance exclusion chromatography on micro-columns packed with LiChrospher silica gel; linear calibration plot," High-Resol. Chromatogr., Chromatogr. Commun., 4, 425-426 (1981).
136. T. Takeuchi, Y. Watanabe, K. Matsuoka, and D. Ishii, "High-back-pressure liquid chromatography. I. Development of micro-high-performance liquid chromatography, using liquefied alkanes as the mobile phase," J. Chromatogr., 216, 153-159 (1981).
137. S. R. Springston and M. Novotny, "Kinetic optimization of capillary supercritical fluid chromatography using carbon dioxide as mobile phase," Chromatographia, 14, 679-684 (1981).
138. M. Novotny, S. R. Springston, P. A. Readen, J. C. Fjelsted, and M. L. Lee, "Capillary supercritical fluid chromatography," Anal. Chem., 53, 407A-414A (1981).
139. P. A. Readen, J. C. Fjelsted, M. L. Lee, S. R. Springston, and M. Novotny, "Instrumental aspects of capillary supercritical fluid chromatography," Anal. Chem., 54, 1090-1093 (1982).
140. Y. Hirata and F. Nakata, "Supercritical fluid chromatography with fused-silica packed columns," J. Chromatogr., 295, 315-323 (1984).
141. T. Takeuchi, D. Ishii, M. Saito, and K. Hibi, "Supercritical fluid chromatography with micro-packed columns and carbon dioxide as a mobile phase," J. Chromatogr., 295, 323-333 (1984).
142. T. Takeuchi and D. Ishii, "High-back-pressure liquid chromatography. III. Open-tubular microcapillary LC using liquefied alkanes as the mobile phase," J. Chromatogr., 240, 51-55 (1982).
143. T. L. Chester, "Capillary supercritical fluid chromatography with flame-ionization detection: Reduction of detection artifacts and extension of detectable molecular weight range," J. Chromatogr., 299, 424-431 (1984).
144. J. C. Giddings, Dynamics of Chromatography, Marcel Dekker, New York (1965).
145. J. H. Knox and M. Salem, "Kinetic conditions for optimum speed and resolution in column chromatography," J. Chromatogr. Sci., 7, 614-622 (1969).
146. C. Horvath and H. J. Lin, "Band spreading in liquid chromatography, general plate height equation, and a method for evaluation of the individual plate height contribution," J. Chromatogr., 149, 43-70 (1978).
147. J. H. Knox, "Practical aspects of LC theory," J. Chromatogr. Sci., 15, 352-369 (1977).
148. K. E. Kaiser and E. Oerlich, Optimization in HPLC, Chromatographic Methods Series, Huthig, Heidelberg (1981).
149. B. G. Belen'kii (Belenky) and L. Z. Vilenchik, Modern Liquid Chromatography of Macromolecules, Journal of Chromatography Library, 25, Elsevier, Amsterdam (1983).

150. P. A. Bristow and J. H. Knox, "Standardization of test conditions for high-performance liquid chromatography columns," Chromatographia, 10, 279-289 (1977).
151. M. J. E. Golay, in: Gas Chromatography 1958, D. H. Desty (ed.), Butterworth, London (1959), pp. 36-84.
152. G. Taylor, "Dispersion of soluble matter in solvent flowing slowly through a tube," Proc. R. Soc. (London), Ser. A, 219, 186-203 (1953).
153. C. A. Kramers, J. A. Rijks, and C. P. M. Schutjes, "Factors determining flow rate in chromatographic columns," Chromatographia, 14, 439-444 (1981).
154. R. Ohmacht and I. Halász, "Properties of commercially available silicas for HPLC," Chromatographia, 14, 155-162 (1981).
155. H. Colin, M. Martin, and G. Guiochon, "Extra-column effects in high-performance liquid chromatography. I. Theoretical study of the injection problem," J. Chromatogr., 185, 79-98 (1979).
156. J. C. Sternberg, in: Advances in Chromatography, Vol. 2, J. C. Giddings and R. A. Keller (eds.), Marcel Dekker, New York (1966), pp. 205-269.
157. Laser Microgel Chromatograph KhZh-1309, Izd. Akad. Nauk SSSR, Nauchno-Teknicheskoe Obshchestvo, Brochure 7581 (1984).
158. New from Waters, Model 680 Automated Gradient Controller, Brochure B 58/July 1982, Waters Associates, Milford (1982).
159. R. P. W. Scott, Liquid Chromatography Detectors, Journal of Chromatography Library, 11, Elsevier, Amsterdam (1977).
160. T. M. Vickrey (ed.), Liquid Chromatography Detectors (Chromatographic Science Series, Vol. 23), Marcel Dekker, New York—Basel (1983).
161. H. Poppe, "Characterization and design of liquid-phase flow-through detector systems," Anal. Chim. Acta, 114, 59-70 (1980).
162. P. C. White, "Recent developments in detection techniques for high-performance liquid chromatography. Part I. Spectroscopic and electrochemical detectors. A review," Analyst, 109, 677-697 (1984).
163. P. C. White, "Recent development in detection technique for high-performance liquid chromatography. Part II. Other detectors. A review," Analyst, 109, 973-984 (1984).
164. P. Kucera, in: Microcolumn High-Performance Liquid Chromatography, P. Kucera (ed.), Journal of Chromatography Library, 28, Elsevier, New York—Amsterdam (1984), pp. 39-73.
165. W. Th. Kok, U. A. Th. Brinkman, R. W. Frei, and H. B. Hanekamp, "Use of conventional instrumentation with micro-bore columns in high-performance liquid chromatography," J. Chromatogr., 237, 357-369 (1982).
166. ISCO Brochure μLC-3D, ISCO, Lincoln, Nebraska, USA, March 1984.
167. T. Takeuchi and D. Ishii, "A multichannel photodiode array ultraviolet—visible detector for micro high-performance liquid chromatography," J. Chromatogr., 288, 451-456 (1984).

168. T. Takeuchi and D. Ishii, "Application of multichannel photodiode array UV—visible detector to the analysis of compounds of a medicine by micro high-performance liquid chromatography," High-Resol. Chromatogr., Chromatogr. Commun., 7, 151-152 (1984).
169. S. V. Kuz'min and N. I. Mikichur, in: Ultramicroanalysis of Nucleic Acids, D. G. Knorre and T. V. Venkstern (eds.), Nauka, Moscow (1973), pp. 133-142.
170. D. Ishii, S. Murata, and T. Takeuchi, "Analysis of bile acids by micro high-performance liquid chromatography with postcolumn enzyme reaction and fluorometric detection," J. Chromatogr., 282, 569-577 (1983).
171. M. Novotny, K. E. Karlsson, M. Konishii, and M. Alassandro, "New biochemical separations using precolumn derivatization and microcolumn liquid chromatography," J. Chromatogr., 292, 159-169 (1984).
172. P. Kucera and H. Umagat, "Design of a post-column fluorescence derivatization system for use with microbore columns," J. Chromatogr., 255, 563-579 (1983).
173. E. S. Yeung and M. J. Sepaniak, "Laser fluorimetric detection in liquid chromatography," Anal. Chem., 1465A-1481A (1980).
174. G. J. Diebold and R. N. Zare, "Laser fluorimetry: Subpicogram detection of aflatoxins using high-pressure liquid chromatography," Science, 196, 1439-1441 (1977).
175. M. J. Sepaniak and E. S. Jeung, "Determination of Adriamycin and Daunorubicin in urine by high-performance liquid chromatography with laser fluorimetric detection," J. Chromatogr., 190, 377-383 (1980).
176. M. L. Aleksandrov, B. G. Belen'kii (Belenky), É. S. Gankina, V. A. Gotlib, J. J. Kever, N. N. Komarov, and V. A. Pavlenko, "Refractometric detector in microcolumn exclusion high-performance chromatography of polymers," High-Resol. Chromatogr., Chromatogr. Commun., 6, 629-632 (1983).
177. J. G. Atwood and M. J. E. Golay, "Dispersion of peaks by short straight open tubes in LC systems," J. Chromatogr., 218, 87-104 (1981).
178. E. D. Katz and R. P. W. Scott, "Low-dispersion connecting tubes for liquid chromatography systems," J. Chromatogr., 268, 169-175 (1983).
179. D. W. Vidrine, Fourier Transform Infrared Spectroscopy, Vol. 2, Academic Press, New York (1979).
180. R. S. Brown, P. G. Amateis, and L. T. Taylor, "Detectability of phenols and amines by normal phase microbore HPLC—FTIR employing highly IR transparent solvents," Chromatographia, 18, (1984).
181. K. Jinno, C. Fujimoto, and D. Ishii, "'Buffer memory' technique for the combination of micro-HPLC and IR spectrometry," J. Chromatogr., 239, 625-628 (1982).
182. C. M. Conroy, P. R. Griffiths, and K. Jinno, "Interface of a microbore high-performance liquid chromatograph with a diffuse reflectance Fourier transform infrared spectrometer," Anal. Chem., 57, 822-825 (1985).

183. D. T. Kuehl and P. R. Griffiths, "Microcomputer-controlled interface between a high-performance liquid chromatograph and a diffuse reflectance infrared Fourier transform spectrometer," Anal. Chem., 52, 1394-1399 (1980).
184. K. Jinno and C. Fujimoto, "Combination of micro-high-performance liquid chromatography and Fourier transform infrared spectrometry using the potassium bromide buffer memory technique," High-Resol. Chromatogr., Chromatogr. Commun., 4, 532-533 (1981).
185. E. J. Caliguri and I. N. Mefford, "Femtogram detection limits for biogenic amines using microbore HPLC with electrochemical detection," Brain Res., 296, 156-159 (1984).
186. S. A. McClintock and W. C. Purdy, "Liquid chromatography/electrochemical detection: A review," Int. Lab., 14, No. 7 (September), 70-80 (1984).
187. S. G. Weber and W. C. Purdy, "Behavior of an electrochemical detector used in liquid chromatography and continuous-flow voltammetry. I. Mass transport limited current," Anal. Chim. Acta, 100, 531-544 (1978).
188. F. A. Posey and R. E. Meyer, "Chronopotentiometry and voltammetry of Ag—AgCl electrode in flowing streams. 2. Theoretical," J. Electroanal. Chem., 30, 359-364 (1971).
189. Y. Hirata, P. T. Lin, M. Novotny, and R. M. Wightman, "Small-volume electrochemical detector for microcolumn liquid chromatography," J. Chromatogr. (Biomed. Appl.), 181, 287-294 (1980).
190. M. Goto, Y. Koyanagi, and D. Ishii, "Electrochemical detector for micro high-performance liquid chromatography and its applications to the determination of aminophenol isomers," J. Chromatogr., 208, 261-268 (1981).
191. K. Šlais and D. Kouřilova, "Electrochemical detector with a 20-nl volume for micro-column liquid chromatography," Chromatographia, 16, 265-266 (1982).
192. K. Šlais and D. Kouřilova, "Minimization of extra-column effects with microbore columns using electrochemical detection," J. Chromatogr., 258, 57-63 (1983).
193. Z. Jin and S. M. Rappaport, "Microbore liquid chromatography with electrochemical detection for determination of nitro-substituted polynuclear aromatic hydrocarbons in diesel soot," Anal. Chem., 55, 1778-1781 (1983).
194. H. B. Hanekamp, P. Bos, and R. W. Frei, "Design and selective application of a dropping mercury-electrode amperometric detector in column liquid chromatography," J. Chromatogr., 186, 489-496 (1979).
195. C. L. Blank, "Dual electrochemical detector for liquid chromatography," J. Chromatogr., 117, 35-46 (1976).
196. M. Goto, T. Nakamura, and D. Ishii, "Micro-high-performance liquid chromatographic system with micro-precolumn and dual electrochemical detector for direct injection analysis of catecholamines in body fluids," J. Chromatogr. (Biomed. Appl.), 226, 33-42 (1981).

197. J. Dutrieu and Y. A. Delmotte, "Dual electrochemical detection in HEFT. New trends in analytical chemistry," Z. Anal. Chem., 314, 416-417 (1983).

198. M. Goto, G. Zou, and D. Ishii, "Determination of catecholamines in human serum by micro high-performance liquid chromatography with micro precolumn and dual electrochemical detection," J. Chromatogr. (Biomed. Appl.), 275, 271-276 (1983).

199. S. G. Weber and W. C. Purdy, "Electrochemical detection with a regenerative flow cell in liquid chromatography," Anal. Chem., 54, 1757-1764 (1982).

200. J. M. DiBussolo, M. W. Dong, and J. R. Gant, "High-speed analysis using electrochemical detection," J. Liq. Chromatogr., 6, 2353-2373 (1983).

201. M. Goto, E. Sakurai, and D. Ishii, "Dual electrochemical detection of biogenic amine metabolites in micro high-performance liquid chromatography," J. Liq. Chromatogr., 6, 1907-1925 (1983).

202. S. Rokushika, Z. Y. Qiu, and H. Hatano, "Micro column ion chromatography with a hollow fiber suppressor," J. Chromatogr., 260, 81-87 (1983).

203. D. Kouřilova, K. Šlais, and M. Krejči, "A conductivity detector for liquid chromatography with a cell volume of 0.1 μl," Collect. Czech. Chem. Commun., 48, 1129-1136 (1983).

204. H. Small and T. E. Miller, Jr., "Indirect photometric chromatography," Anal. Chem., 54, 462-469 (1982).

205. S. Rokushika, Z. Y. Qiu, Z. L. Sun, and H. Hatano, "Microbore packed-column anion chromatography using a UV detector," J. Chromatogr., 280, 69-76 (1983).

206. H. D. Lux and E. Neher, "The equilibrium time course of $[K^+]_0$ in cat cortex," Exp. Brain Res., 17, 190-205 (1973).

207. E. Ujec, E. E. O. Keller, N. Kriz, V. Pavlik, and J. Machek, "Low-impedance, coaxial, ion-selective, double-barrel microelectrodes and their use in biological measurements," Bioelectrochem. Bioenerg., 7, 363-369 (1980).

208. T. Tsuda, A. Nago, G. Hakawaga, and M. Maseki, "Adaptation of moving-wire flame ionization detector to accept total effluent in micro-liquid chromatography," High-Resol. Chromatogr., Chromatogr. Commun., 6, 694-695 (1983).

209. E. P. Lankmayr, H. J. Hayes, B. L. Karger, P. Vorous, and Y. M. McGuire, "Band broadening phenomena with the moving belt LC—MS interface," Int. J. Mass Spectrom. Ion Phys., 46, 177-180 (1983).

210. V. V. Brazhnikov, L. M. Yakushina, E. M. Syanova, and S. A. Karapetyan, "Application of electron-capture detector to the testing of high-performance liquid chromatography microcolumns," High-Resol. Chromatogr., Chromatogr. Commun., 6, 451-453 (1983).

211. U. A. Th. Brinkman, R. B. Geerdink, and A. DeKok, "Normal- and reversed-phase liquid chromatography with on-line electron-capture detection," J. Chromatogr., 291, 195-201 (1984).

212. V. L. McGuffin and M. Novotny, "Thermionic detection in microcolumn liquid chromatography," Anal. Chem., 55, 2296-2302 (1983).
213. O. Von Stetten and R. Schlett, "Purification of ^{125}I labeled compounds by high-performance liquid chromatography with online detection," J. Chromatogr., 254, 229-235 (1983).
214. G. Schwedt, Chemische Reaktionsdetektoren für die Schnelle Flüssigkeits-Chromatographie, Hüthig, Heidelberg (1980).
215. R. W. Frei and A. H. M. T. Scholten, "Reaction detectors in HPLC," J. Chromatogr. Sci., 17, 152-160 (1979).
216. W. B. Farmon, Continuous Flow Analysis. Theory and Practice, Marcel Dekker, New York (1976).
217. R. W. Frei, in: Chemical Derivatization in Analytical Chemistry, Vol. 1, Chromatography, R. W. Frei and J. F. Lawrence (eds.), Plenum Press, New York (1981).
218. Sj. Van der Wal, "Post column reaction detection systems in HPLC," J. Liq. Chromatogr., 6, Suppl. 1, 37-59 (1983).
219. P. Kucera and H. Umagat, in: Microcolumn High-Performance Liquid Chromatography, P. Kucera (ed.), Journal of Chromatography Library, 28, Elsevier, Amsterdam (1984), pp. 154-178.
220. J. A. Apffel, U. A. Th. Brinkman, and R. W. Frei, "Use of nonsegmented flow, post-column reaction detection with miniaturized HPLC systems," Chromatographia, 17, 125-131 (1983).
221. N. G. Anderson, R. H. Stevens, and J. W. Holleman, "Analytical techniques for cell fractions. X. High-pressure ninhydrin reaction system," Anal. Biochem., 26, 104-117 (1968).
222. M. L. Aleksandrov, B. G. Belen'kii, N. N. Komarov, V. G. Mal'tsev, V. B. Melas, and N. N. Sudareva, "Optimization of amino acid detection conditions by ninhydrin reaction," Zh. Prikl. Khim. (Leningrad), 56, 2097-2100 (1983).
223. M. A. Berezhkovskii, B. G. Belen'kii, R. G. Vinogradova, M. B. Ganitskii, N. E. Zhiltsova, V. G. Mal'tsev, and O. P. Shilov, in: Instruments for Scientific Investigations and Experiment Automation, V. A. Pavlenko (ed.), Nauka, Leningrad (1982), pp. 121-127.
224. T. Takeuchi and D. Ishii, "Analysis of bile acids in serum by micro HPLC," High-Resol. Chromatogr., Chromatogr. Commun., 6, 571-572 (1983).
225. T. Takeuchi, S. Saito, and D. Ishii, "Micro high-performance liquid chromatographic separation of bile acids," J. Chromatogr., 258, 125-134 (1983).
226. A. Hirose, Y. Iwasaki, K. Ueda, and D. Ishii, "Post column colourimetric detection with xylene orange in micro-HPLC of rare earth metals," High-Resol. Chromatogr., Chromatogr. Commun., 4, 530-531 (1981).
227. J. A. Apffel, U. A. Th. Brinkman, and R. W. Frei, "Design and application of a post-column extraction system compatible with miniaturized liquid chromatography," Chromatographia, 18, 5-10 (1984).
228. C. E. Buffett and M. D. Morris, "Microcell thermal lens detector for liquid chromatography," Anal. Chem., 55, 376-378 (1983).

229. K. Jinno and S. Nakanishi, "Micro-HPLC-IP detection system in separation of non-UV absorbing organic compounds," High-Resol. Chromatogr., Chromatogr. Commun., 6, 210-211 (1983).
230. K. Jinno, H. Zsuchida, S. Nakanishi, Y. Hirata, and Ch. Fujimoto, "Micro-high-performance liquid chromatography—inductively coupled plasma combination technique in analysis of organometallic compounds," Appl. Spectrosc., 37, 258-261 (1983).
231. K. Jinno, S. Nakanishi, and T. Nagoshi, "Microcolumn gel permeation chromatography with inductively coupled plasma emission spectrometric detection," Anal. Chem., 56, 1977-1979 (1984).
232. R. C. Massey, C. Crews, D. McWeeny, and M. E. Knowles, "Analysis of a model ionic nitrosamine by microbore high-performance liquid chromatography using a thermal energy analyzer chemiluminescence detector," J. Chromatogr., 236, 527-529 (1982).
233. M. Krejči, K. Šlais, and K. Tesařik, "Electrokinetic detection in liquid chromatography measurement of the streaming current generated on analytical and capillary columns," J. Chromatogr., 149, 645-652 (1978).
234. M. Krejči, in: Instruments for Scientific Investigations, V. A. Pavlenko (ed.), Materials of International Scientific and Technical Conference of the Member Countries of the Council for Mutual Economic Aid on Scientific Instruments — Nauchpribor-78, SÉV, Moscow (1980), pp. 359-363.
235. A. Stolyhwo, H. Colin, and G. Guiochon, "Use of light scattering as a detector principle in liquid chromatography," J. Chromatogr., 265, 1-18 (1983).
236. A. Stolyhwo, H. Colin, H. Martin, and G. Guiochon, "Study of the qualitative and quantitative properties of the light scattering detector," J. Chromatogr., 288, 253-275 (1984).
237. D. R. Bobbit and E. S. Yeung, "Absorption detection in microcolumn liquid chromatography via indirect polarimetry," Anal. Chem., 57, 271-274 (1985).
238. K. Hiyaguchi, K. Honda, and K. Imai, "Microbore HPLC and chemiluminescence detection of DNS amino acids," J. Chromatogr., 316, 501-506 (1985).
239. R. B. Green, "Lasers: Practical detectors for chromatography?" Anal. Chem., 55, 20A-32A (1983).
240. J. M. Charlesworth, "Evaporative analyzer as a mass detector for liquid chromatography," Anal. Chem., 50, 1414-1420 (1978).
241. R. P. Arpino, G. Guiochon, P. Krien, and G. DeVant, "Optimization of the instrumental parameters of a combined liquid chromatograph—mass spectrometer coupled by an interface for direct liquid introduction. I. Performance of the vacuum equipment," J. Chromatogr., 185, 529-547 (1979).
242. G. Guiochon and P. J. Arpino, "How to interface a chromatographic column to a mass spectrometer," J. Chromatogr., 271, 13-25 (1983).
243. D. E. Games and E. D. Ramsey, "High-performance liquid chromatography—mass spectrometry of derivatized and underivatized amino acids," J. Chromatogr., 323, 67-70 (1985).

244. R. G. Christensen, H. S. Hertz, S. Meiselman, and E. White, "Liquid chromatograph—mass spectrometer interface with continuous sample preconcentration," Anal. Chem., 53, 171-174 (1981).
245. D. E. Games, N. J. Alcock, L. Cobelli, C. Eckers, M. P. L. Games, A. Jones, M. S. Lant, M. A. McDowall, M. Rossiter, R. A. Smith, S. A. Westwood, and H. Y. Wong, "LC/MC studies with moving belt interfaces," Int. J. Mass Spectrom. Ion Phys., 46, 181-184 (1983).
246. P. B. Arpino, M. A. Baldwin, and F. W. McLafferty, "Liquid chromatography—mass spectrometry. 2. Continuous monitoring," Biomed. Mass Spectrom., 1, 80-82 (1974).
247. P. R. Jones and S. K. Yang, "Liquid chromatograph/mass spectrometer interface," Anal. Chem., 47, 1000-1003 (1975).
248. E. C. Horning, D. I. Carrol, I. Dzidic, K. D. Haegele, M. G. Horning, and R. N. Stillwell, "Liquid chromatograph—mass spectrometer computer analytical systems. Continuous flow system based on atmospheric pressure ionization mass spectrometry," J. Chromatogr., 99, 13-17 (1974).
249. M. L. Vestal, "Studies of ionization mechanisms involved in thermospray LC—MS," Int. J. Mass Spectrom. Ion Phys., 46, 193-196 (1983).
250. P. Dobberstein, E. Korte, G. Meyerhoff, and R. Pesch, "Investigation of an LC/MS interface for EI, CI, and FAB ionization," Int. J. Mass Spectrom. Ion Phys., 46, 185-188 (1983).
251. H. Jungclas, H. Danigel, and L. Schmidt, "Liquid chromatography/mass spectrometry, with ^{252}Cf fission fragment induced ionization," Int. J. Mass Spectrom. Ion Phys., 46, 197-200 (1983).
252. M. L. Aleksandrov, L. N. Gal, N. V. Krasnov, V. I. Nikolayev, V. A. Pavlenko, and V. A. Shurov, "Atmospheric pressure ion extraction from the solutions — a new method of mass-spectrometric analysis of bioorganic substances," Dokl. Akad. Nauk SSSR, 277, 379-383 (1984).
253. R. D. Smith, J. Fjeldsted, and M. L. Lee, "Supercritical fluid chromatography—mass spectrometry," Int. J. Mass Spectrom. Ion Phys., 46, 217-220 (1983).
254. P. J. Arpino and G. Guiochon, "LC—MS coupling," Anal. Chem., 51, 682A-701A (1979).
255. J. Henion, in: Microcolumn High-Performance Liquid Chromatography, P. Kucera (ed.), Journal of Chromatography Library, 28, Elsevier, Amsterdam (1984), pp. 260-298.
256. A. P. Bruins, "Developments in interfacing microbore high-performance liquid chromatography with mass spectrometry. A review," J. Chromatogr., 323, 99-111 (1985).
257. B. L. Karger and P. Vouros, "A chromatographic perspective of HPLC—MS," J. Chromatogr., 323, 13-32 (1985).
258. C. R. Blakley and M. L. Vestal, "Thermospray interface for liquid chromatography/mass spectrometry," Anal. Chem., 55, 750-754 (1983).

259. F. J. Yang, in: Proceedings of the Fourth International Symposium on Capillary Chromatography, R. E. Kaiser (ed.), Institute of Chromatography, Bad Dürkheim, Heidelberg (1981), pp. 139-176.

260. F. J. Yang, "Open-tubular column LC: Theory and practice," J. Chromatogr. Sci., 20, 241-251 (1982).

261. T. Takeuchi, D. Ishii, and A. Nakanishi, "Instrumentation for fast micro high-performance liquid chromatography," J. Chromatogr., 285, 97-102 (1984).

262. T. Takeuchi and D. Ishii, "Valve injection in micro-HPLC," High-Resol. Chromatogr., Chromatogr. Commun., 4, 469-470 (1981).

263. P. Kucera and G. Guiochon, "Use of open-tubular columns in liquid chromatography," J. Chromatogr., 283, 1-20 (1984).

264. K. Šlais, D. Kouřilova, and M. Krejči, "Trace analysis by peak compression sampling of a large sample volume on microbore columns in liquid chromatography," J. Chromatogr., 282, 363-370 (1983).

265. H. Schwartz and R. G. Brownlee, "A dual syringe LC solvent delivery system for use with microbore columns," Int. Lab., 14, No. 9 (November/December), 38-52 (1984).

266. J. Sjodahl, H. Lundin, R. Erikson, and J. Ericson, "Solvent delivery in the new application areas of high-performance liquid chromatography," Chromatographia, 16, 325-329 (1982).

267. M. Martin and G. Guiochon, in: Instrumentation for High-Performance Liquid Chromatography, J. F. K. Huber (ed.), Journal of Chromatography Library, 13, Elsevier, Amsterdam—New York (1978), pp. 41-85.

268. H. A. H. Billiet, P. D. M. Keehnen, and L. DeGalan, "Single-pump solvent programmer for high-performance liquid chromatography using synchronized valve switching," J. Chromatogr., 185, 515-528 (1979).

269. H. A. Quarry, R. L. Grob, and L. R. Snyder, "Measurement and use of retention data from high-performance gradient elution contributions from 'non-ideal' gradient equipment," J. Chromatogr., 285, 1-18 (1984).

270. C. R. Powley, W. A. Howard, and L. B. Rogers, "Mixing considerations in the development of a gradient microbore high-performance liquid chromatographic system," J. Chromatogr., 299, 43-57 (1984).

271. R. J. Perchalsky and B. Wilder, "Reversed-phase liquid chromatography at increased temperature," Anal. Chem., 51, 774-776 (1979).

272. K. Jinno and Y. Hirata, "Investigation of the low temperature effect in micro high-performance liquid chromatography — open tubular microcapillary column," High-Resol. Chromatogr., Chromatogr. Commun., 4, 466-468 (1981).

273. Y. Hirata and K. Jinno, "High-resolution liquid chromatography with a packed micro glass capillary column," High-Resol. Chromatogr., Chromatogr. Commun., 6, 196-199 (1983).

274. I. Halász and E. E. Heine, "Separation of low-boiling hydrocarbons by gas chromatography using packed capillary columns," Nature (London), 194, 971-979 (1962).
275. K. P. Hupe, R. J. Jonker, and G. Rosing, "Determination of band-spreading effects in high-performance liquid chromatographic instruments," J. Chromatogr., 285, 253-265 (1984).
276. K. W. Freebairn and J. H. Knox, "Dispersion measurements on conventional and miniaturized HPLC systems," Chromatographia, 19, 37-47 (1984).
277. L. R. Snyder and J. J. Kirkland, Introduction to Modern Liquid Chromatography, Wiley, New York (1979), p. 215.
278. J. P. Wolf, "Large-diameter columns for preparative-scale high-speed liquid chromatography," Anal. Chem., 45, 1248-1250 (1973).
279. A. B. Littlewood, in: Gas Chromatography 1964, A. Goldup (ed.), Institute of Petroleum, London (1965), pp. 77-82.
280. C. E. Schwartz and J. M. Smith, "Flow distribution in packed beds," Ind. Eng. Chem., 45, 1209-1218 (1953).
281. J. C. Giddings, "Generalized nonequilibrium theory of plate height in large-scale gas chromatography," J. Gas Chromatogr., 1, 38-42 (1963).
282. C. L. DeLigny and W. E. Hammers, "Diffusive and convective dispersion in chromatography, recent developments," J. Chromatogr., 141, 91-105 (1977).
283. C. H. Eon, "Comparison of broadening patterns in regular and radially compressed large-diameter columns," J. Chromatogr., 149, 29-42 (1978).
284. J. H. Knox and J. F. Parcher, "Effect of column-to-particle-diameter ratio on the dispersion of unsorbed solutes in chromatography," Anal. Chem., 41, 1599-1606 (1969).
285. J. H. Knox, G. R. Laird, and P. A. Raven, "Interaction of radial and axial dispersion in liquid chromatography in relation to the 'infinite diameter effect'," J. Chromatogr., 122, 129-145 (1976).
286. R. G. Avery and J. D. F. Ramsay, "Sorption of nitrogen in porous compacts of silica and zirconia powders," J. Colloid Interface Sci., 42, 597-603 (1973).
287. J. Asshauer and I. Halász, "Reproducibility and efficiency of columns packed with 10 μm silica in liquid chromatography," J. Chromatogr. Sci., 12, 139-147 (1974).
288. M. Verzele, J. Van Dijck, P. Mussche, and C. Dewaele, "Spherical versus irregular-shaped silica gel particles in HPLC," J. Liq. Chromatogr., 5, 1431-1448 (1982).
289. M. Verzele, "Miniaturization of the particle size and low dispersion liquid chromatography. Low viscosity solvent upward packing procedure," J. Chromatogr., 295, 81-87 (1984).
290. M. Broguaire, "Simple method of packing HPLC columns with high reproducibility," J. Chromatogr., 170, 43-52 (1979).
291. H. A. Claesseus, G. Aben, and N. Vonk, "Packing procedures of silica columns for HPLC with aqueous slurries," High-Resol. Chromatogr., Chromatogr. Commun., 5, 250-253 (1982).

292. T. J. N. Webber and E. H. McKerrel, "Optimization of LC performance on columns packed with microparticulate silicas," J. Chromatogr., 122, 243-258 (1976).
293. R. M. Cassidy, D. S. LeGay, and R. W. Frei, "Study of packing techniques for small-particle silica gels in high-speed liquid chromatography," Anal. Chem., 46, 340-344 (1974).
294. J. L. Medina and S. Dave, "How to pack a high-efficiency column," Int. Lab., 14 (March), 90-91 (1984).
295. K. Kuwata, M. Uebori, and Y. Yamazaki, "Rapid method for packing microparticulate columns packed with chemically bonded stationary phase for HPLC," J. Chromatogr., 211, 378-382 (1981).
296. R. F. Meyer and R. A. Hartwick, "Efficient packing of small particle microbore columns," Anal. Chem., 56, 2211-2214 (1984).
297. H. G. Menet, P. C. Gareil, and R. H. Rosset, "Experimental achievement of one million theoretical plates with microbore liquid chromatographic columns," Anal. Chem., 56, 1770-1773 (1984).
298. R. Majors, "High-performance liquid chromatography on small particle silica gel," Anal. Chem., 44, 1722-1726 (1972).
299. Sh. A. Karapetyan, L. M. Yakushina, G. G. Vasiyarov, and V. V. Brazhnikov, "Slurry packing of high-performance columns for liquid chromatography: Part 1: Effect of quality of tube inner wall," High-Resol. Chromatogr., Chromatogr. Commun., 6, 440-441 (1983).
300. N. K. Vadukul and C. R. Loscombe, "A comparative study of microbore reversed-phase columns," 5, 360-363 (1982).
301. P. Welling, H. Poppe, and J. C. Kraak, "Preparation of efficient and stable reversed-phase microbore columns for high-performance liquid chromatography," J. Chromatogr., 321, 450-457 (1985).
302. N. K. Vadukul and C. R. Loscombe, "An investigation of microbore reversed-phase columns containing 3 μm packing materials," High-Resol. Chromatogr., Chromatogr. Commun., 6, 488-491 (1983).
303. R. Ohmacht and I. Halász, "Efficiency of commercially available silicas in HPLC," Chromatographia, 14, 216-226 (1981).
304. F. J. Onuska, M. E. Comba, T. Bistricki, and R. T. Wilkinson, "Preparation of surface-modified wide-bore wall coated open tubular columns," J. Chromatogr., 142, 117-125 (1977).
305. J. D. Schieke, N. R. Comins, and V. Pretorius, "Whisker-walled open-tubular glass columns for gas chromatography. Techniques for inducing whisker growth," J. Chromatogr., 112, 97-107 (1975).
306. D. C. Locke, J. T. Schermund, and B. Banner, "Bonded stationary phases for chromatography," Anal. Chem., 44, 90-102 (1972).
307. I. Sebestian and I. Halász, "Monomere Chemisch Gebundene Stationarephasen fur die Gas und Flüssigkeits-Chromatographie mit Si—C Binding," Chromatographia, 7, 371-375 (1974).
308. O. E. Brust, I. Sebestian, and I. Halász, "Si—N bonded stationary phases for liquid chromatography," J. Chromatogr., 83, 15-24 (1973).

309. R. K. Iler, The Chemistry of Silica, Wiley-Interscience, New York (1979), pp. 872-877.
310. T. Tsuda and M. Novotny, "Band broadening phenomena in microcapillary tubes under the conditions of liquid chromatography," Anal. Chem., 50, 632-634 (1978).
311. A. Manz, Z. Fröbe, and W. Simon, in: Microcolumn Separations, M. V. Novotny and D. Ishii (eds.), Journal of Chromatography Library, 30, Elsevier, Amsterdam (1985), pp. 297-307.
312. J. S. Landy, J. L. Ward, and J. G. Dorsey, "A critical evaluation of some stainless steel and radially compressed reversed-phase HPLC columns," J. Chromatogr. Sci., 21, 49-56 (1983).
313. S. Terabe, K. Otsuka, and T. Ando, "Electrokinetic chromatography with micellar solution and open-tubular capillary," Anal. Chem., 57, 834-841 (1985).
314. T. Tsuda, K. Nomura, and G. Nakagawa, "Open-tubular microcapillary LC with electro-osmosis flow using a UV detector," J. Chromatogr., 248, 241-245 (1982).
315. M. Novotny, in: Microcolumn High-Performance Liquid Chromatography, P. Kucera (ed.), Journal of Chromatography Library, 28, Elsevier, Amsterdam (1984), pp. 194-259.
316. J. C. Giddings, M. N. Myers, L. McLaren, and R. A. Keller, "High-pressure gas chromatography of nonvolatile species. Compressed gas is used to cause migration of intractable solutes," Science, 162, 67-73 (1968).
317. E. Klesper, A. H. Corwin, and D. A. Turner, "Porphyrin studies. XX. High-pressure gas chromatography above critical temperatures," J. Org. Chem., 27, 700-709 (1962).
318. M. S. Vigdergauz, A. V. Garasov, V. A. Ezrets, and V. I. Siomkin, Gas Chromatography with Nonideal Eluents, Nauka, Moscow (1980), pp. 130-131.
319. G. J. Schmidt, D. C. Olson, and W. Slavin, "Amino acid profiling of protein hydrolyzates using liquid chromatography and fluorescence detection," J. Liquid Chromatogr., 2, 1031-1045 (1979).
320. T. Takeuchi, M. Yamazaki, and D. Ishii, "Micro HPLC of 5-dimethylaminonaphthalene sulphonylamino acids," J. Chromatogr., 295, 333-341 (1984).
321. T. Takeuchi, M. Yamazaki, and D. Ishii, "Isocratic separation of PTH amino acids by micro HPLC," High-Resol. Chromatogr., Chromatogr. Commun., 7, 101-102 (1984).
322. R. L. Cunico, R. Simpson, L. Correia, and C. T. Wehr, "High-sensitivity PTH-amino acid analysis using conventional and microbore chromatography," J. Chromatogr. (Biomed. Appl.), 336, 105-113 (1984).
323. D. H. Spackman, W. H. Stein, and S. Moore, "Automatic recording apparatus for use in the chromatography of amino acids," Anal. Chem., 30, 1190-1206 (1958).
324. F. E. Regnier, "High-performance liquid chromatography of biopolymers," Science, 222, 245-252 (1983).

325. É. S. Gankina, I. O. Kostyuk, and B. G. Belen'kii, "High-efficiency gel chromatography of large peptides," Bioorg. Khim., 5, 325-326 (1979).
326. I. O. Kostyuk, É. S. Gankina, and B. G. Belen'kii, "Microcolumn chromatography of large peptides," Bioorg. Khim., 8, 1047-1051 (1982).
327. É. S. Gankina, I. O. Kostyuk, and B. G. Belen'kii, "Microcolumn chromatography of large peptides," in: New Approaches in Liquid Chromatography, H. Kalász (ed.), Akadémiai Kiadó, Budapest (1984), pp. 231-242.
328. H. Wilgus and E. Stelimagen, "Calibration mixture for estimation of peptide molecular weight by exclusion chromatography," Anal. Biochem., 94, 228-230 (1979).
329. E. C. Nice, C. J. Lloyd, and A. W. Burgess, "The role of short microbore HPLC columns for protein separation and trace enrichment," J. Chromatogr., 296, 153-170 (1984).
330. M. J. O'Hare, M. W. Capp, E. C. Nice, N. H. C. Cooke, and B. G. Archer, "Factors influencing chromatography of proteins on short alkylsilane-bonded large pore-size silicas," Anal. Biochem., 126, 17-28 (1982).
331. M. W. Hunkapiller, R. M. Hewick, W. J. Dreyer, and L. E. Hood, in: Methods in Enzymology, Vol. 91, C. H. W. Hirsand and S. N. Timusheff (eds.), Academic Press, New York—London (1983), pp. 399-413.
332. G. K. Ackers, in: Advances in Protein Chemistry, Vol. 24, C. B. Anfinsen, M. L. Anson, J. T. Edsalland, and F. M. Richards (eds.), Academic Press, New York—London (1969), pp. 343-344.
333. L. M. Nichol and D. J. Winsor, Migration of Interacting Systems, Clarendon Press, Oxford (1972).
334. G. H. Ackers, in: Methods in Enzymology, Vol. 27, C. H. W. Hirs, N. Serge, and S. H. Timasheff (eds.), Academic Press, New York (1973), pp. 441-445.
335. V. G. Mal'tsev, T. M. Zimina, O. I. Kurenbin, B. G. Belen'kii, M. L. Aleksandrov, N. P. Pavlova, V. L. Dyakov, and V. K. Antonov, "Microchromatographic studies of the kinetics and equilibrium of association of the phospholipase A_2 from the venom of middle Asian cobra," Bioorg. Khim., 5, 1710-1719 (1979).
336. V. G. Mal'tsev, O. I. Kurenbin, T. M. Zimina, L. Z. Vilenchik, Yu. Ya. Gotlib, and B. G. Belen'kii, "Microchromatographic studies of the association of a protein interacting with the sorbent," Bioorg. Khim., 6, 1053-1061 (1980).
337. V. G. Mal'tsev, T. M. Zimina, B. G. Belen'kii, O. I. Kurenbin, N.P. Pavlova, V. P. Dyakov, and V. K. Antonov, "Microchromatographic studies of the association of the phospholipase A_2," Bioorg. Khim., 8, 96-101 (1982).
338. V. G. Mal'tsev, T. M. Ziminia, B. G. Belen'kii, and O. I. Kurenbin, "Microbore column exclusion chromatographic method for studying protein association and its relation with enzymatic activity," J. Chromatogr. (Biomed. Appl.), 273, 95-105 (1983).

339. K. Jinno and M. Nishihara, "Application to gel permeation chromatography of micro-high-performance liquid chromatography," Anal. Lett., 13, 673-681 (1980).
340. J. J. Kever, B. G. Belen'kii, É. S. Gankina, L. Z. Vilenchik, O. I. Kurenbin, and T. P. Zmakina, "Determination of molecular-weight distribution of polymers by microcolumn exclusion chromatography," Vysokomol. Soedin., B24, 403-406 (1982).
341. L. Z. Vilenchik, O. I. Kurenbin, T. P. Zmakina, and B. G. Belen'kii, "The choice of a sorbent with linear calibrating dependence in the gel permeation chromatography of polymers," Vysokomol. Soedin., A22, 2801-2804 (1980).
342. B. G. Belen'kii and L. Z. Vilenchik, in: Chromatography of Polymers, Khimiya, Moscow (1978), p. 223.
343. L. Z. Vilenchik, O. I. Kurenbin, T. P. Zmakina, V. V. Nesterov, E. V. Chubarova, and B. G. Belen'kii, "A new method of chromatography correction of the instrumental band broadening," Vysokomol. Soedin., A22, 2804-2809 (1980).
344. T. Takeuchi, D. Ishii, and S. Mori, "Separation of oligomers by high-performance micro gel permeation chromatography," J. Chromatogr., 257, 327-335 (1983).
345. T. Takeuchi and D. Ishii, "Gradient separation of complex mixtures by micro high-performance liquid chromatography," High-Resol. Chromatogr., Chromatogr. Commun., 6, 310-317 (1983).
346. E. R. White and D. N. Laufer, "Reversed-phase HPLC of antibiotics on microbore columns," J. Chromatogr., 290, 187-197 (1984).
347. K. Tsuji and R. B. Binns, "Micro-bore HPLC for the analysis of pharmaceutical compounds," J. Chromatogr. (Biomed. Appl.), 253, 227-237 (1982).
348. Y. Fujii, H. Fukuda, Y. Saito, and M. Yamazaki, "Separation of digitalis glycosides by micro HPLC," J. Chromatogr., 202, 139-143 (1980).
349. Y. Fujii, H. Fujii, and M. Yamazaki, "Separation and determination of cardiac glycosides in Digitalis purpurea leaves by micro HPLC," J. Chromatogr., 258, 147-153 (1983).
350. P. Kucera and R. A. Hartwick, in: Micro-Column High-Performance Liquid Chromatography, P. Kucera (ed.), Journal of Chromatography Library, 28, Elsevier, Amsterdam (1984), pp. 179-183.
351. S. Rokushika, Z. Y. Qui, and H. Hatano, "Microbore packed-column anion-exchange liquid chromatography of nucleo-bases, nucleosides, and nucleotides," J. Chromatogr., 320, 335-342 (1985).
352. A. N. Wulfson and S. A. Yakimov, "HPLC of nucleotides. II. General methods and their development for analysis and preparative separation. An approach to selectivity control," High-Resol. Chromatogr., Chromatogr. Commun., 7, 442-460 (1984).
353. R. Raydrick, A. Terragno, and R. Tackett, "Picogram detection of eicosanoids by ultraviolet absorbance after narrow-bore HPLC. Comparison with conventional-bore columns," J. Chromatogr. (Biomed. Appl.), 308, 31-41 (1984).

354. I. R. Shipe, A. E. Arlinghaus, J. Savory, M. K. Wills, and J. P. Dimaro, "Determination of bethanidine in plasma by liquid chromatography with micro-bore reversed-phase column," Clin. Chem., 29, 1793-1795 (1983).
355. Y. Fujii, R. Oguri, A. Matsuhashi, and M. Yamazaki, "Micro HPLC separation of 3,5-dinitrobenzoyl derivatives of cardiac glycosides and their metabolites," J. Chromatogr. Sci., 21, 495-499 (1983).
356. M. Novotny, M. Alassandro, and M. Konishi, "Microcolumn liquid chromatography of benzoyl derivatives of steroid metabolites," Anal. Chem., 55, 2375-2378 (1983).
357. C. H. Lochmüller, M. L. Hunnicutt, and R. W. Beaver, "Separation of nitro-polycyclic aromatics on bonded-pyrene stationary phases in microbore columns," J. Chromatogr. Sci., 21, 444-446 (1983).
358. K. Jinno, "High-speed micro-high-performance chromatography for analysis of polycyclic aromatic hydrocarbons in environmental samples," High-Resol. Chromatogr., Chromatogr. Commun., 7, 218-219 (1984).
359. A. Nakanishi, "Determination of antioxidants in gasoline by micro high-performance liquid chromatography," J. Chromatogr., 291, 398-403 (1984).
360. A. Hirose, D. Wiester, and M. Novotny, "High-efficiency microcolumn separation of polycyclic arene isomers isolated from carbon black," Chromatographia, 18, 1239-1242 (1984).
361. R. C. Debeise, W. Rieman, and S. Lindenbaum, "Analysis of halide mixtures by ion-exchange chromatography," Anal. Chem., 26, 1840-1841 (1954).
362. G. Muto, Y. Takata, and H. Tsuda, "Ion-exchange chromatography of halide mixtures in mixed solvents," Nipon Kagaku Zasshi (J. Chem. Soc., Jpn.), 88, 432-435 (1967).
363. J. N. Story and J. J. Fritz, "Forced-flow chromatography of the lanthanides with continuous in-stream detection," Talanta, 21, 892-894 (1974).
364. D. H. Sisson and V. A. Mode, "High-speed separation of the rare earth by ion exchange, Part II," J. Chromatogr., 66, 129-135 (1972).
365. T. Hayashi and T. Yanabe, "Elution behavior of rare earth elements on single and mixed ion-exchange columns," J. Chromatogr., 87, 227-231 (1973).
366. Y. Tokata and Y. Arikawa, "Cation-exchange chromatography of rare-earth elements," Bunseki Kagaku (Jpn. Analyst), 24, 762-767 (1975).
367. T. Tsuda, T. Nozu, and G. Nakagawa, "Separation of chloro complexes of metals by micro-high-performance liquid chromatography," J. Chromatogr., 242, 331-336 (1983).
368. M. Cooke, "Determination of nitrate in water by microbore HPLC," High-Resol. Chromatogr., Chromatogr. Commun., 6, 383-386 (1983).
369. M. V. Novotny and D. I. Ishii (eds.), Microcolumn Separations, Journal of Chromatography Library, Vol. 30, Elsevier, Amsterdam (1985).